Teaching Science through the Grades

Compiled and edited by
DAVID MITCHELL

If a child is to keep alive his inborn sense of wonder without any such gift from the fairies, he needs the companionship of at least one adult who can share it, rediscovering with him the joy, excitement and mystery of the world we live in.

– Rachel Carson

Printed with support from the Waldorf Curriculum Fund

Published by:

Waldorf Publications at the
Research Institute for Waldorf Education
351 Fairview Avenue
Suite 625
Hudson, NY 12534

Title: *Teaching Science through the Grades*
Editor: David Mitchell
Principal translator: Ted Warren
Layout: Ann Erwin
© 2007 by AWSNA Publications, reprinted 2020
ISBN #978-1-943582-42-6

Gratitude is expressed to the editors of *Steinerskolen* and the individual authors for granting permissions to translate the essays for North America, to the editors of *Steiner Education*, and to Peter Glasby and Neil Carter, editors of the *Journal for Waldorf Education* in Australia and New Zealand.

Contents

5 Foreword

7 A Study of the Element "Water"
Christian Smit

19 Water as the Medium for Life
Jørgen Smit

26 Goethe's *Theory of Colors*
Tørger Holtsmark

30 Zoology and Mythology
Jens Bjorneboe

35 Chemistry in Grades Seven to Nine
Jan Haakonson

52 Astronomy: The Oft-Forgotten School Subject
Sven Bohn

64 The Starry Heavens and Our Self
Jørgen Smit

70 Teaching Biology in a Human Context
Graham Kennish

77 Aesthetic Knowledge as a Source for the Main Lesson
Peter Guttenhöfer

Contents

84 Adolescents—Their Relationship to the Night and the Senses in Connection with Their Own Development
Peter Glasby

104 Thoughts on Information and Communication Technology
Florian Oswald

Foreword

These articles were originally published as part of the *Waldorf Journal Project*, a series sponsored by the Waldorf Curriculum Fund to bring translations of essays, magazine articles, and specialized studies from around the world to English-speaking audiences. This series is now being republished in book format. This ninth edition of translations is comprised of articles intended to strengthen science teaching in the elementary grades and take us inside the thought life of Waldorf teachers from Norway. Four lectures follow, one on biology from Great Britain, one from Germany, one from Switzerland, and one from Australia/New Zealand that focus on major thoughts of both our time and our striving as Waldorf teachers.

The first two articles written by the Smit brothers, Jørgen and Christian, explore the nature and phenomena of water. Tørger Holtsmark, a professor of physics from the University of Oslo (who was one of my mentors when I taught in Norway) explores Goethe's *Theory of Colors*. International award-winning novelist/poet/playwright (and Waldorf teacher) Jens Bjorneboe develops the connection between zoology and mythology. Chemistry from grades 7–9 is the focus of the next article by Jan Haakonson. Sven Bohn establishes astronomy as an important middle school subject, and Jørgen Smit returns to connect the outer world with the inner world with a simplicity with which he is unique. Graham Kennish from Great Britain carries this theme forward from the perspective of biology. The final two essays deal with overarching themes proposed by Rudolf Steiner. Peter Guttenhöfer and Peter Glasby explore aesthetic knowledge and the trinity of conclusion, judgment and concept. In the concluding chapter, Florian Oswald from Berne, Switzerland, shares his thoughts on communication and information technologies.

We hope that this collection will help teachers and others gain insight into the seeds of Waldorf education. The editor, David Mitchell, was interested in receiving your comments on the material for this publication during his lifetime.

We at Waldorf Publications are interested, in the spirit of his limitless interest in research from the field and in the news, in hearing from you. We would also be interested in hearing what areas you would like to see represented in future *Journal* projects. If you know of specific articles that you would like to see translated, please contact Waldorf Publications.

> – Patrice Maynard
> Waldorf Publications
> at the Research Institute for Waldorf Education

A Study of the Element "Water"

by Christian Smit
translated by Ted Warren

In middle school physics, it is fruitful to bring well-known phenomena into more clear and conscious light. Such an everyday topic is water. Because we take water for granted and we assume we know so much about it, we rarely reflect on the element's being and meaning. Every child enjoys splashing in it. No one outgrows the fun of water. But they may be surprised to discover the nature of this element.

We studied water in a three-week block using our imagination, curiosity and flexible thinking. We practiced sculpting thoughts and concepts in order to follow the winding stream on which water led us.

The water cycle as the basis for life

This theme is traditionally taught in the fifth grade biology lessons so we began our study with review: The warmth of sunlight draws water from the seas, oceans, moist earth and vegetation up into the atmosphere. There the water forms a layer of wet air around the earth which condenses into rain, snow and hail that wander back to the earth, seas and oceans. We recall that water-movements in fresh water begin in the different levels in rivers, waterfalls and rapids, while movements in salt water begin with temperature differences, wind pressure, the rotation of the earth, and tides.

Thunderstorm rolling onto land from the ocean

We admire the miraculous balance between evaporation and the discharge of rivers into the oceans during this pulsating cycle. We look for other ways in which water or any fluid moves through a cycle. Our attention falls upon the great ocean currents, and we make drawings of the Gulf Stream before focusing on the circulation of blood in our bodies. We agree that the movement of fluids provides a unique foundation for life.

We took a detour to the sea floor and the various underwater formations. On the edge of every ocean there are common formations: the continental depression—the egg that falls down, the enormous, deep ocean with relatively few deep-sea cavities.

To follow the water cycle, children need to think in a way that does not lock them in to finite details, but rather enables them to flow with the descriptions of the development of related phenomena.

Water's tendency to the globular form

Our next step was to find a forming tendency of water; the one most readily apparent is the water drop. All solid bodies have the ability to maintain their structure outwardly. This structure can occur through accident (breakage), or it can be defined geometrically by internal forces that are expressed in crystallization. Solid bodies have countless shapes, but water demonstrates only one common form, namely the globe. All liquid materials have one common dimension; they can only build one surface form, namely the globular form.

In beautiful water drops that shine while falling, we see the globular tendency. Yet we also see the potential for water to collect in a round globe in huge formations of the oceans of the world. Also solid materials do as water—they join in the globular form. All erosion and disintegration tend to create globes. The fact that materials pack together from

Water droplets on a rose bush. Note the reversed reflective image within the drop.

every direction and build a globe appears as primeval phenomena in the shape of a water drop. These observations result in what we normally call gravitation. Phenomenally, we find all of the parts of physical materials trying to collect in a common globular form.

When we look at hail, we find the globular tendency, which can be further seen in layers of globular rinds one on top of the other. It is easy to experience every lake, pond and puddle as parts of global surfaces which in their totality would cover the earth in concentric layers. This is convincingly demonstrated in communication pipes. Water seeks the same level in every pipe that means the global surface is parallel to the ocean surface. It does not matter how thick or thin the pipes are. The amount of water in the differently shaped pipes will hold each other in balance because, as coherent amounts of water, they will tend to a common globular form.

At this point the pupils were asked to describe the relationship between floating liquids and the tendency to form globes. An eighth grader described our experiences in these words: "Water's tendency to form globes appears in drops, which you will see if you pour water on a warm plate, or water on impregnated clothes, and so forth. If the water is not allowed to take its own globular form, it will collect parallel to the surface of the earth, which is itself an enormous globe."

A water level is also a globe or, more correctly, as a part of a globe. If you build a house anywhere on the surface of the earth, the level will follow the surface of the earth. Gravity causes this effect. It collects everything into a middle point. But fluids have the ability to pull back into themselves. Because of gravity they can form globes, within and of themselves. Conclusion: globular tendency = gravity.

We can say that solid bodies are individualized in their ability to pull together—they individualize gravity—while liquids collect in a unified gravitational form. In other words, liquids, when they are as small as a drop, can break free from the power of gravity and create their own world. This moved our study on to surface tensions, but before we worked with that phenomena, we observed how water moves and forms itself in rivers, rapids, waterfalls and waves.

Water tends to move in zigzags and whirlpools. We all know that rivers form zigzags on their way to the sea. We see it most clearly when rivers slow down in flatlands. The winding zigzags can be so strong that they form small lakes adjacent to the river. As the river water flows through the zigzags, it makes a spiral movement both in the outer banks and the inner banks of the river.

We demonstrated the spiral movement of the water by pouring a flat, broad stream of water from a milk carton. A beautiful spiral forms. Water wants to wind around itself. We have seen this many times before, and the pupils enjoy this familiar phenomenon in a new, existential context.

Meandering stream

We used a simple apparatus to demonstrate the water spiral in a glass container with a hole in the bottom where a rubber tube is attached so that the speed of water emission is controlled. When the water is released, it forms the same spiral as we see in the bath tub. Only now we can view the spiral from the side. The tip of the spiral "hangs" at the spout, while the whirlpool forms an upward funnel while swinging from side to side. If the water flow is stopped, the tip of the spiral is lifted up from the vessel; when the water is released again, the spiral tip sinks down to the hole. The whole process reminds us of a miniature tornado. It is a fascinating and beautiful play!

We reflect upon the fact that water never moves in a straight line, but in a spiral form. Once again our thoughts have softened up and allow us to feel with our senses. If it is allowed to play on its own terms, water cannot be held to straight lines and formations. This enlivens our observation and thinking.

Form springs from movement

The next day we try to remember, as well as possible, any waterfalls we have seen. Water streams off the rock formations and with a majestic expression thunders down toward cliffs and stones. The spray rises and white, splashing formations fill the landscape in foaming water at the bottom of the falls. What is going on?

We observe a fairly constant form in the waterfall and in the water below. What is taking place? The movement of the water creates a constant form while the materials stream more or less unnoticed through the form. It is a majestic performance of a law of life. The material that is formed through movement is subordinate and exchangeable. We discuss this observation and then look at our own bodies as examples. After seven years practically all of the materials in our bodies have been exchanged for new materials, but the form remains.

In the waterfall the water funnel is exchanged in a short period of time; in our body's waterfall the process takes much longer. The form of our body is not determined by a collection of separate parts, but it is a movement that is filled with material in rhythmical exchange.

Let us observe the waves of the ocean. This brings us to a new phenomenon. The forward moving wave travels through water that stands almost still in its place. A water particle in a wave moves in a circle at the surface. The movement is more elliptic under the water and approaches a back and forth movement farther down. A stormy ocean is a complicated combination of unlimited rows of movement that grasp the same materials at the same place. There may be only one body in each place on land, but in water there can be different movements in the same place that culminate in a final, sensible expression in a certain form only to be exchanged in the next moment for a new one. Flexible thinking is required to follow the game. A good exercise is to reverse the experiences. First review the observation of a waterfall, where the movement creates a constant form that is continually streaming with new material. Next consider waves in the ocean, where the form moves in a forward motion while the material remains in the same place. These are opposite systems that both express the moving element: water.

Surface tension

We return to detailed observations of the shape of water, the surface of the globe. Water takes the shape of the vessel it is in. The surface of water is parallel to the earth. A certain amount of isolated water can create a globe from its own power, for example a water drop that falls. Any fluid will contract to a globe if it is able to form freely. This tendency is clearly demonstrated in the case of mercury that approaches the shape of a globe even when in its most solid form.

12 *A Study of the Element "Water"*

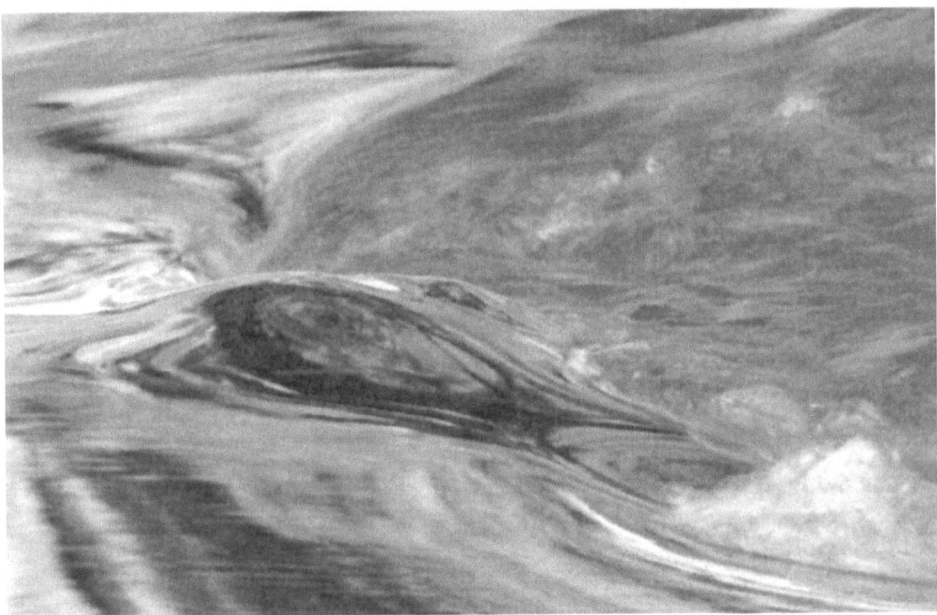

Running brook showing surface tension

What are the special geometric qualities of a globe? Of all of the platonic bodies, the globe has the least surface in relation to its volume. This means that any liquid can approach the globe shape by making its surface area as small as possible. The inner necessity for a liquid to make its surface as little as possible can be demonstrated through a number of experiments and thereby enlightens our study of the concept of surface tensions.

A glass vessel filled with water is set before the class. A beautiful squirrel tail with bushy hair is swung in the air a couple of times before it is dipped into the water. When we pull it up, that hair naturally clings together, and the tail looks small and miserable. I ask the pupils, "Why has the squirrel tail stuck together?"

The answer is usually, "Because the tail is wet." To demonstrate that this answer is not enough, I dip the tail once again in the water and hold it there. What do we see? The hair flows back and forth in a bushy formation under the water. The tail is wet, but the hair does not cling together. Once again we pull the tail out of the water and the hairs stick together again. A membrane, a "layer of skin" of water is created around the tail hair. Before our eyes a contraction along the surface takes place. We experience the concept of surface tension.

Another experiment is to carefully place a razor blade in the water. It floats. We weight it down with half a match and add six more. We notice that the blade has made an indention in the surface of the water. Another visible experiment can be carried out with a finely netted wire basket filled with wax. The wires are coated in wax but the holes are open. A sizable wire net can float on water but if you touch it, it quickly sinks to the bottom.

Water's tendency to make as small a surface as possible can be demonstrated in other ways as well. Each pupil makes a ring of thin steel thread. In their ring they loosely tie together some thin threads to make a mask. When this steel-thread ring together with the thin threads is dipped into soapy water, a membrane will form over the entire form, and the threads will float in the soapy membrane. When we put a hole in one of the masks, it will immediately take the shape of a circle. Stick a hole in the neighboring mask and it will immediately make a circle, but where the two circles connect, a straight line will be formed. The hole grows as much as possible because the soapy membrane around it tries to contract together to the smallest surface. The circle expands geometrically into the largest surface in relation to the periphery. The remaining masks that are still covered by the water membrane will have reached their absolute minimum. We observe the surface tension pull all membranes into a minimum. This same "compression tendency" drives free-flowing water and all other liquids to create the smallest spatial form, the globe.

Water contracts on the outside into a "skin." Where this skin appears, the water approaches a more solid form, creating something that resembles a solid border surface. The solid border surface is a permanent quality all solid substances have without effort. Water and all liquids recreate this skin continuously, and the skin dissolves as soon as the water returns to its own element. The skin appears only in relation to another element; it cannot appear in its own element.

Our experiments become even more complicated when we go beyond working with a membrane in a flat circle on one plane to using wire-meshings in a three-dimensional figure. I pass out thin wire thread to the pupils, who each make a spatial figure. We start with regular figures such as pyramids, cubes or cones. In every case the surface tension will create the minimum surface. A soapy water membrane will cling to the wires everywhere but also create surfaces that hang together in the holes. A cube made of wire thread will have eight slightly bent-over surfaces inside with a square that holds them together in the middle. With a little practice you can create an inner cube of water membrane materials that hang elegantly in slightly bent-over membranes, held to the thread on its side. The inner water cube contains six membranes with air inside, held together merely by water membranes.

At this point we test the pupils' inventive abilities. Some make the most complicated, irregular forms in order to see what will happen, while others experiment to observe the laws of freely moving water figures. The whole time they can be surprised by the forms created by the pure liquids. There was a lot of variation: symmetrical, asymmetrical, bowed-over, distorted shapes that were made of steel thread by the children.

Capillaries—water's tendency toward gas formations

We notice that water at the top of the thin tube does not arch upward but pulls upward along the sides, creating an arched valley. "But that cannot be true!" cry out the excited pupils. "That goes against everything we've learned about water pulling together to form a globe!" Now we are in the middle of an exciting problem and must expand our understanding of water.

We start new experiments by (1) dipping a piece of sugar in colored water, and (2) dripping ink on absorbing paper. We watch how the liquid in both cases spreads into the material. It is absorbed over the surface. Every pupil is given a flame and a thin glass tube of 20–30 cm. The glass tube is warmed up in the middle until it is soft. Quickly and decisively we pull it out to a length of 60–80 cm, thus a capillary tube with minimal inner sections. We break off one side to the point where the capillary tube shows that water moves in thin canals, and with this experience we look again into water cycles in vegetation. Water can climb up 50–60 meter tall trees. And what happens on top? There the water evaporates into the atmosphere. Capillaries may be considered a stage in water's path to evaporation. In the earth, water collects in pools by the roots; through the trunk, branches, twigs, and leaves, it is thinned out by increased capillary movement until it escapes out into the air as water-steam in gas form.

This is the opposite phenomenon to the tendency to form globes as water tries to approach solid materials. In the capillary movements water tries to approach gases. "Where else do we find capillaries?" I ask the children. In bogs or other places, water is sifted into more solid materials. A bog can also be found in the lymph system of the human body. Our whole body is filtered by water, and without that hydrating of our body, we are not worth much. Once again we touch on water's life-bearing activities. The ability to spread out, climb up, thin out in all directions—these are tendencies in capillaries that are qualities of gases. Gases do this on their own, but water needs a partner to support its becoming more and more like gas with evaporation and centripetal spreading.

Here a further experiment can help to demonstrate capillary movement. In a vessel with colored water we place a rag that reaches from the water over the edge and down to another vessel alongside. The liquid will slowly penetrate the rag over to the other vessel. This is known as osmosis. Another demonstration can be created with two glass plates put together with a matchstick to keep them apart. A rubber band holds them together. These plates are placed in a colored liquid and the liquid will be pulled up in a beautiful arc along the side where the glass plates are placed together.

Pressure in water

We fill a vessel with water that is connected by a rubber hose at the bottom to a fountain. The higher we raise the vessel, the higher the spout of water will flow out of the fountain. We realize that the difference in height will affect the speed with which the water leaves the fountain. Pressure increases with the difference in height. Then we connect the vessel with a little container that has holes from all of its corners. When we raise the vessel with water, the waterspout will stream from every corner. No matter how we hold the container, the stream of water spreads to every corner. We see that pressure in water moves to every side and is strongest where the distance up to the water surface in the vessel is largest.

To make it even more obvious, we place a messing pipe on the highest edge. Along the pipe are holes bored in a row. We fill the pipe with water and the waterspout is at the bottom, where the pressure is greatest, and continually shorter toward the top, where the pressure is least. Again, we see that the water pressure increases with height.

Now we can speak about the relationship between weight and pressure. We ask if water has any weight at all. Can it be weighed? Not without being placed in a container. You always have to fill water into a container to judge whether there is a weight increase in relation to the emptied container.

We use this experience to ask, "What happens when we place an empty container in water?" The container becomes lighter because it becomes buoyant. Weight and buoyancy are opposite symmetrical phenomena and may be explained: "Water in a body adds weight; water surrounding a body reduces weight."

We can take a step further and say: If you prevent a stone from falling, you realize its weight. If you prevent water from flowing freely, you notice its pressure. The weight of solid bodies is analogous to flowing medium's horizontal surface. Just as weight can only appear within a confined surface, so must buoyancy have limited surface to appear. A natural conclusion is that amounts of water that are down in water and surrounded by water float in total balance. It has neither "weight" nor "buoyancy," nor does it have a surface where pressure may occur. We can say: Water surrounded by water is weightless, and not subject to gravity.

We try to avoid presenting conclusions. Instead, the concepts our pupils build about water need to be based on experiences they have with water. This allows the children to remain open and flexible in their thinking. Their enthusiasm for the subject grows as their thoughts become more connected and enlivened. This is our educational goal with the lessons on water.

The hydrostatic paradox

We demonstrate the hydrostatic paradox, or Pascal's tube as it is also called. Three tubes of different shapes are mounted on a scale. One tube is straight and is the same width along its whole length. A second tube expands as it moves toward the opening at the top, while the third tube becomes smaller and is more lopsided. Each tube has a round, circular opening at the bottom that is covered by a thin rubber membrane.

When the tubes are mounted by weight, the indicator moves to the same point as soon as water in the tube has reached a certain height. This is the same for all three tubes. The result is dependent on the volume of water in the pipe, on the pressure. The students can understand that the same result appears in the tube that expands toward the top and in the straight tube because they can imagine that the water rests on the lopsided edges. The fact that the result is the same for water in the tight, lopsided tube awakens second thoughts and excitement. The question arises: "How can there be almost no 'weight' of water in the small tube, and we still get the same result?"

Let us summarize all of our experiences with the three main phenomena of water:
1. The tendency for water to form a globe with a skin-like substance and surface tension and an approach to the solid, formation building
2. Capillary action, siphoning, hydrating, osmosis, evaporation, movement
3. Pressure that searches for balance and that takes the form of globe surface

Vortice train created by drawing a feather through water on which has been sprinkled lycopodium powder from the spore cases of club mosses.

We have experienced how necessary water is for life, how strongly it forms, but also how it continually removes the created forms if no solid substance enters and allows itself to be formed by the movement. The formative forces in water appear beautifully in streaming whirlpools. These may also be seen when an object moves through water, for example, when we row on a summer morning on a mirror-perfect sea. We can also create these conditions in the classroom. A vessel 30 x 60 cm and only a few centimeters deep will work well. Paint the vessel black inside; fill it with water that has glycerin added to it in order to make the water a little harder. Spread lycopodium powder over the water surface and move a stick through the liquid. Beautiful whirlpools will appear and quickly disappear. This leads us to the philosophical statement, "All solid bodies are movement that has become still."

Water's life-bringing element is the basis for movement and formation, but not for the ability to let the formations become solid. To become solid, another element is needed. This we will explore in the geology block. Solids are the remains of movement. The study of the element water leads us to new studies in nature.

Christian Smit was a class teacher at the RudolfSteinerskole in Oslo and a founding teacher of the RudolfSteinerskole in Bærum.

Water as the Medium for Life

by Jørgen Smit
translated by Ted Warren

If we observe a tree trunk, for example an old, rugged oak trunk, we observe something that was once alive. Through the tree trunk, life still streams. But in the form before us, life has almost completely disappeared. As soon as the living parts of the plant become hard and stiff, the plant is removed from the living stream. In the forms created by the dying substance, we observe a life-stream in a congealed condition. The dead or partially dead plant materials still serve life as a supportive base, and are thereby part of the totality of the living organism. New life can grow forth on the old, half-dead trunk's solid formations. But where does it grow forth? Only there where something is still soft enough that there is a possibility for a streaming, rhythmical movement. It needs no more than a weak indication of such streams, but the possibility must be there.

The prerequisite for such streaming movements to appear is liquid. The streaming, moving can also be air, for example in our lungs, but for the most part flowing liquids are the element of life. Solids are either a supportive base for the streaming, moving liquids or they are actually in a dissolved condition when taken into the moving life-stream. Of all of the liquids, water has a unique ability to enable life to unfold.

Water is a liquid that needs most warmth to become warm itself and then holds warmth the longest. This fact has a decisive meaning for all life on earth. If we observe the earth in its totality, we discover that the oceans take in the warmth from the tropical zones. In the streams of the oceans, the warmth is circulated into polar regions, reducing the cold. Without the Gulf Stream Norway's climate would be much like Greenland's. If we think hypothetically of another liquid in the great oceans, a liquid that would heat up easily and therefore also easily give off heat—if this were the case, then the great balance between the tropical regions and the polar regions would not take place, and all life on earth would soon disappear.

The same balance is also found in the atmosphere's streaming water steam. Normally a liquid's specific weight increases when it becomes colder, and the coldest layers sink. This is also true of water, but only down to roughly +4º Celsius. Below that temperature, water becomes lighter and the cold climbs upward. Therefore ice always begins to form on the surface of water and not at the bottom. Warmth is thereby held through the winter more successfully to a larger extent than if the ice formation were to begin on the bottom. Without this special quality of water the whole earth would be quickly transformed into a golden desert, for the most part an ice-desert. But thanks to the element of water, life has an opportunity to unfold.

Spring ice encapsulating a shoot of a sapling

A very special water quality plays a huge role in the organism for human life. The human body's temperature ideal temperature is +37º Celsius (98.6ºF). A little variance either way means sickness, life-threatening conditions or even death. That temperature +37º Celsius is the optimum temperature for water to become a little warmer. At this point the organisms of life and water's qualities resound harmoniously together.

How does water relate to the solid substances? We find a special condition that enables water to optimally serve life. From the perspective of chemistry, water is neutral. But among the neutral liquids water has the highest ability to dissolve solid substances and to hold solid substances in dissolved or suspended conditions. Living organisms need many solid substances for nutrition. But first the solids must be dissolved or suspended in water.

Therefore when we try to better understand how life unfolds, we must continually observe water, especially the great water cycles over the entire planet. The water cycle is modified by daily rhythms and by the rhythms of the seasons. All too often we are tempted to observe a plant in an isolated situation. It has the streaming water within, but the specific qualities and movements of the streaming water are related to all of the daily and seasonal rhythms on the earth.

In a fast-running creek, water takes on countless forms that change continually. These forms are closely related to the organic forms, especially in plant life. It is very difficult to hold a single streaming water form tightly in our mental images. That would also be unfair to the creek, for it does not stand still. But a stiffened result of the streaming activities can be observed easily in winter when the creek is frozen with ice. The unstoppable, formation movement is gone. In the swollen formations we can observe a stiffened picture of the "living" water movements.

Now it would be a great mistake to identify life as streaming water. Life is much more than streaming water. But the streaming water is a friend of life and a servant of life. It is life's necessary medium.

When water begins to stream, an inner movement structure in the liquid arises immediately. The speed is greatest in the middle and less on the sides where water is slowed by friction. Let us now assume that a part of the water in the middle has a certain speed and the side parts have a slower speed. Two borders will arise at the place of transition. Quickly speed differences will arise in each part, also new places of transition, and so forth. Speed in flowing water therefore becomes gradually, steadily more rapid from the sides toward the middle, and the water formations that glide by each other can be as thin as the walls of soap bubbles. Therefore enormous surfaces arise. In every cubic centimeter of streaming water hundreds of cubic-meters of inner water surfaces can be formed.

The surfaces of these forms remain at the great surface tensions (which are also special water qualities). If a certain speed is reached, whirlpools arise and the wondrous, "living," streaming formations of falling stream.

Let us now consider the amount of water in a falling stream. We may ask: How much does it weigh? The answer is: 1.2 kilograms (2.65 pounds). What is the chemistry formula? Answer: H_2O. In many life-situations it is the questions are the most important, and answers to these questions do not always tell us the most important aspects of streaming water. The weight of water is just a factor in the game. In the minute water takes on streaming movements, where the enormous inner surfaces arise in continuous formation transformation, it reaches more than ever the "field of life." I reference this reality intentionally by using the word *field*.

Now we can describe some experiments performed by Theodor Schwenk in his laboratories at the Weleda factory in Schwabisch Gmund, just outside of Stuttgart, Germany. Mr. Schwenk was an engineer who specialized in the study of the streams we find in water and air.

The first phase

A number of glass bottles of the same shape are filled ⅔ with distilled water. At 7:00am the first bottle is shaken for four minutes and returned to its place with a label stating the time. At 7:15am the next bottle is shaken. All of the other bottles are shaken, one every fifteen minutes. Finally he has 96 (24 x 4) bottles standing in his laboratory. The only difference between the bottles of distilled water is that they were shaken at different times.

The second phase

The next day he takes 96 new bowls and writes the 96 different times on them. He places a bit of wheat seed in each of them. (Biodynamic wheat seeds are best for this experiment.) Then he pours the distilled water from each of the 96 bottles into the corresponding bowl without allowing a minute's pause between each. The bowls sit for two days. In each bowl there are many seeds that begin to sprout, and some seeds that do not sprout.

The third phase

The following day he takes 96 new bowls and labels them with the time of each shaking in the first phase. Each bowl is filled with tap water. From each bowl in phase two he chooses 30 seeds that are all sprouting and places them group-wise in the corresponding bowl with tap water. Only the seeds with sprouts are used in the remainder of the experiment. Now he has 96 bowls and in each of them are 30 wheat seeds that have begun to sprout. He lets them grow in the tap water bowls for six days. The temperature, pressure, moisture and light are the same for every bowl in the room.

The fourth phase

By the end of the sixth day, the sprouts have grown to approximately 10 cm in height. He measures the height of all of the 30 sprouts in a bowl and finds the average height. The same is done for each of the other bowls and groups of sprouts. It turns out that the 96 averages vary with up to 1 cm difference. He draws a curve where the original time of the shaking of the water is one factor and the average height of the sprouts is another factor. (A control is also done to weigh all of the sprouts. The average weight for the 96 groups is calculated; the result is the same as the height measurements.)

Had he merely conducted this experiment and been able to draw only one curve, the result would have already been very obvious. When he has calculated the average height of the 30 sprouts for each bowl (and all of the non-sprouting seeds are removed), one should assume that the 96 average heights would be the same. There would be a slight possibility for the 30 seeds that had the greatest growth intensity (that appears in the seed's height and weight) accidentally ended up in the same bowl. The difference is too large to consider a coincidence as the explanation.

Meanwhile Schwenk and his coworkers have conducted not only one but also a whole series of these experiments: this same experiment conducted practically every day for seven years. In other words 2000 experiments were made, resulting in 2000 curves. The curve's maximum and minimum became clearer and clearer the more curves were drawn. The experiments indicated clearly that the sprout's growth-intensity (measured by height of the sprout and the weight after a determined time) is different according to the corresponding seed sprouting in distilled water that was rhythmically shaken at different times. If the distilled water was shaken at 2:00pm, it received a different quality as the medium for growth intensity than the distilled water that was shaken, for example, at 3:30pm. This must relate to the earth's daily rhythm. During every day the atmospheric conditions change (light, warmth, moisture, and so forth) in a cycle from sunrise to noon, sunset and midnight. Experiments show that water, when shaken rhythmically at a certain time, brings in a greater "field" and thereby has new qualities according to the field's condition. With the rhythmical shaking, water enters a streaming movement. The enormous inner surfaces create a continual and never-ending form transformation. Therefore water becomes "sensible" for activities of the greater "field." It becomes a more powerful medium for life.

Schwenk's experiments show that there is a relationship that can be further explored. In his book, *Grundlagen der Potenzforschung* (Weleda–Verlag, Schwabisch Gmund, 1954), Schwenk explains these experiments that provide a basis for new experiments on the "potencies" of materials.

Let us indicate some of these, iron sulfate for example. Schwenk used a 1% solution of iron-sulfate. He mixed an amount of it in a ten-fold amount of water and mixed the new mixture rhythmically. We call the first potency (D1). We take one-tenth of the amount and mix it with a ten-fold amount of water, rhythmically. This is potency (D2). We continue this way until we mix the thirtieth potency of iron-sulfate solution (D30). As the amount of iron-sulfate is reduced, the potency number is higher. After the twenty-fourth potency, one could assume that, according to normal molecule theory, we would not have a single iron-sulfate molecule in the mixture, as the tenth in the twenty-fourth potency is a higher number than the number of molecules. Potencies over twenty-four are therefore considered identical with pure water. Further, we should expect a gradually reduced level of activity from D1 to D24. The experiments, however, show that this is not the case.

Schwenk carried out the above-mentioned series of experiments with a variation; to the water in which the seeds sprouted, he added a solution of iron-sulfate in different potencies from D1 to D30. The growth intensity curves he recorded do not match the corresponding curve for distilled water. Clear differences, even for potencies above the twenty-fourth potency, appear. The differences between these curves and the water curves are equally reduced from D1 to D30. Three peaks of activity appear: the first, with the lowest grade of potency; second, with the twelfth potency; and third, with the twentieth level of potency. And between these three maximums are two decisive areas of minimum activity.

We can conclude that, when the amount of iron sulfate is reduced, new activities appear that are clearly not bound to the amount of materials. The reduction is not indifferent. If we throw a handful of iron-sulfate in the North

Sea, eventually an equally strong dilution will take place as with higher potencies. But the new activities that are bound to the amount of materials do not appear. They appear only as dilution gradually takes place and especially every time with rhythmical shaking, where the enormous inner surfaces appear in continual form transformation. (This is the case for potencies using water as the medium.) With the results of many experiments that are now carried out, it is unwise to deny the existence of these activities that are not bound to the original amount of materials.

Yet this is a new area of research where the work has just begun. It is a very long and precipitous journey from proving the existence of an activity to learning how to control and use it, for example in medicine. Some research is already published. Other experiments have not been made public yet but are based on solid research. We can look forward to the further publications from Schwenk and other experimental laboratories.

Jørgen Smit was a class teacher at the RudolfSteinerskole in Bergen, a founding teacher of the Rudolf Steinerseminariet in Järna, Sweden, and for many years a member of the Vorstand in Dornach, Switzerland.

Goethe's *Theory of Colors*

by Tørger Holtsmark
translated by Ted Warren

Goethe's *Theory of Colors* is one of the least accessible documents in the history of modern interpretations of nature. The writer considered it his main piece of work and his testament. At the same time he compared himself to a chess player who had merely made his introductory moves. He knew many years would pass before the general public would understand his distant goal. With bitter irony he stated that his theory would rest in a dormant state until the year 2000. What is the position it now holds in 2007?

A renaissance for Goethe's *Theory of Colors* has not taken place, but many outstanding scientists have considered it necessary to take a position to it. From the highest, most responsible academic quarters it has been stated that Goethe's method holds the seed to a new approach to nature and that this approach is more encompassing than the natural scientific approach we have today. They add that Goethe's science is yet a distant, future possibility and that mankind must continue to follow its present course to the end.

One thing is certain: From an educational perspective Goethe's *Theory of Colors* is an important and highly relevant document. Obviously one can never teach it as it is. That would be a misunderstanding, for his actual presentation was limited by the knowledge available at the time and his conflict with Newton was merely of local, historical interest. One aspect of the theory that we can learn a great deal from is that colors are treated as objective realities in nature. Nor is the traditional border drawn between so-called subjective and objective sensory qualities.

We can better understand the questions involved if we consider for a moment, Goethe's "original phenomena" within the theory of colors, which state that yellow and blue are the first colors; they are the qualities of the "partially transparent medium" in its relation to light and darkness. Through the atmosphere the universe tones toward us in the color blue, and according to the thickness of the air, it takes on every grade of blue until it goes over to black-violet on the high mountain tops. On the other hand, the sun represents light; it tones in yellow, and, according to the thickness of the air, it takes on yellow-red values. On a cloudy, fall day the sun is ruby red.

We know Goethe's "original phenomena" from other relationships, for example smoke. From a Newtonian point of view, the "original phenomena" are explained by the different wavelengths that are absorbed and spread to various degrees in the atmosphere. Goethe uses another language. He speaks of the, "partially transparent medium," allowing the media to be the carrier of color in the same way that tones are carried by air. Goethe places the same objective interest on colors as Pythagoras placed on the sound relationships of a monochord. Sound may also be reduced to vibrations in a medium, namely air. But for Goethe and Pythagoras sound and color are qualities of a corresponding medium, namely air and "the partially transparent medium." The qualities of sound and color can be defined just as we otherwise define the qualities of weight and movement. Everyone can research the physical conditions of Goethe's and Pythagoras' qualities. Pythagoras found them in the monochord's physical dimensions; Goethe found them in the relationship between light and darkness.

The educational value of Goethe's *Theory of Colors* is that it opens a qualitative approach to optical phenomena. Let us take our study of Goethe's theory further and see what are its valuable aspects beyond the purely educational. How is light represented in his *Theory of Colors*? Goethe said that "colors are the deeds of the light" and that light itself is invisible. Light is also a quality, according to Goethe, but a quality of the higher order. Light is active in nature in the same way that a painter speaks about light in his paintings. Goethe's experience of nature is close to an artist's. For an artist color is a quality, and his practical problem is to allow the quality to appear. Quality is always the expression of its own activity, certain obstinacy you might say. Artistic colors are not dead materials but active materials.

Goethe observed nature with the eye of an artist. He saw the same things as his counterpart, Newton. He even carried out many of the same experiments as Newton, but their methods of contemplation and their overall goals were different. Newton resolved all of his experiments to the quantitative, physically measurable results, and so we have spectral colors. Goethe's experiments were for the qualitative aspects, for process, and so we have the color hexagon and the study of contrasts, such as light and darkness.

In nature the same activities take place according to strict and eternal laws, as they take place in every piece of art that is created. For Goethe art was of a higher nature. The artist allows a higher totality to speak through the qualitative polarities. In this way Goethe observed in nature a main principle that he defined as "polarity and elevation." Every natural process or event is a qualitative change, a metamorphosis. When we say that a change is "natural," we mean that it is not coincidental, but follows a characteristic, typical pattern. In metamorphosis, a change between opposite qualities, a higher totality always appears. Goethe called this idea or representation at times an "original phenomenon." In his way the artist does the same as nature. But while nature's cycle is limited, its representation eternal and unchangeable, the artist, in contrast, is creative; he brings forth something new, something no one else has seen, though his materials are the same as nature's.

Let us observe how the Goethean principles of nature appear in his *Theory of Colors*. The original colors, the main aspects of the medium, are yellow and blue. A change between yellow and blue takes place, a metamorphosis, or an elevation as we defined above. The elevation of yellow and blue takes place through all of the variations of yellow-red, orange, and cinnabar on one side and blue-red, violet, and red violet on the other side. But on the divide between the reddish-yellow and the reddish-blue values sits a color in which Goethe was very interested. It is purple, a bright red color that resolves the polarity between yellow and blue. Words can barely describe purple. It is often called "the color of peach blossoms." The question is whether purple is realized in nature and as a pigment color, or whether it is a theme that nature plays upon, a goal for which it strives. Purple is the zenith of Goethe's world of colors.

But there is another transition between yellow and blue, namely green. It is not created by an elevation, but rather by a reduction, in that we can speak of two components, yellow and blue, and their mutual dominance in the color green. According to Goethe, color's variety can be summed up in a symbolic, geometric scheme, his color hexagon. Orange, violet and green are representatives for the

Johann Wolfgang von Goethe 1749–1832

three transitions in the kingdom of color, while yellow, blue and crimson present the primary qualities. It is curious that Newton, due to his method of contemplation, focused on the transitional colors. Here Newton found his measurable sizes and spectral colors while Goethe's colors of yellow, red and blue were considered by Newton as mixtures of spectral colors. In a strange way Goethe's and Newton's opinions stand diametrically opposed, and we must learn to understand that both theories are correct in their own ways and we can accept both opinions. No doubt within Goethe's comprehensive materials there is not a single error that a Newtonian scientist can point to, and within Newton's comprehensive experiences there is no phenomena that Goethe did not come to know and consider in his own way.

Newton's doctrine has not been silent for four hundred years. It has seen significant practical consequences all this time, color photography and color television to name just two. His work has had not only a logical worth but also practical applications. What practical worth has Goethe's *Theory of Colors* had? The technical applications, as far as I know, have as yet been insignificant, but its potential for education and human understanding is profound, for it casts light on the relationship between art and science. Goethe claimed that he who would follow his observations of nature would achieve inner freedom.

Tørger Holtsmark is a professor of physics at the University of Oslo and a long-time advisor to the Waldorf teachers in Scandinavia.

Zoology and Mythology

by Jens Bjorneboe
translated by Ted Warren

Zoology is actually an aspect of mythology. Many of the most well known animal species appeal to such powerful feelings that they provoke the same, deep layer of our unconscious that can be touched only with the most simple and true myths. The word *tiger* contains just as mystical and secret a sound as the word *Cain*. *Eagle*, *lamb*, *lion* and *hare* are all words that can penetrate right through us and sink into deep layers of our soul, like a stone that falls far into the earth through a deep mine-shaft.

Elephant—what an endless, soul-like firmness lies in that word! And what a remarkable group of sounds, vowels and consonants meet each other here! *Nightingale*, *swan* and *shark*! *Butterfly* and *rhinoceros*! Yes, there may be no doubt that we belong to their family, together with lonesome *wolves*, mother *hens* and rooster *chicks*. We are created in their picture also. In our distant past we must have had a lot in common.

The great English mystic and author, William Blake, wrote the beautiful poem on the tiger, and it begins with two lines that are already powerful:

> Tiger, tiger, burning bright,
> In the forest of the night...

He describes the tiger as a mystical, supernatural being, the true tiger of fairy tale-like dimensions, liberated from nature's most awesome secrets.

Lessons in zoology build upon the dream-like dimensions of the animal, on the soul of the animal and the most complete equivalent, its ideal physical expression. Every animal is a human characteristic, a soul condition held in eternal form.

What is a shark?

The shark is made of teeth. It is mouth and tail. It roams through miles of ocean with open mouth, eating everything from anchor-chains to octopi, and the teeth grow so quickly that they spew out of the mouth, so that shark-filled waters have sea floors filled with shark teeth and sand. On the other hand, the shark has almost no digestion; everything it swallows, it remits half-digested. Its intestines are worthless; just a short stump that sucks in the required nutrition; then it empties its stomach and continues its eternal hunt—because the stomach cannot hold on to any food. This is how the shark plows through the ocean, crazed for hunger, ready to swallow anything. With practically no brain, it is almost insensitive to pain other than hunger; it can ignore wounds and knife lacerations.

In reality the shark is an unchanged species; ages ago it achieved its form and its unbeatable specialty, the naked combination of movement and hunger; teeth and muscle. It is the most primitive fisherman we know, specialized at an early point in time. The shark is the embodiment of the fall of man, solidified and unchanged. It fell out of evolution, a pair of lonely jaws eternally moving through the oceans with no other goal than that of reducing its nagging, never-ending hunger—this is the soul of the shark.

What a difference compared with the little, soft silk-ape with large, lively eyes, and the quick little heart that hammers away under his fur coat, warm and silent.

When the peregrine falcon attacks, it nosedives toward the bird it will take, usually one of the good fliers, at a speed of 300 kilometers per hour. With one claw it hits the bird, not over the back or in the neck, but with surgical accuracy it inserts a claw through the cranium of the bird, in the middle of the crown of the head in a lightning fast trepanning. In the hands of a talented falconer, an enlightened falconer, the peregrine falcon becomes identical with its archetype, with human thought. Man and bird become one being, they live together day and night, until the man becomes the bird and the bird becomes the man. In other words it becomes a part of him that it has always symbolized—the thought, free as a bird—and the bird, free as the thought.

He who has observed hens and roosters without discovering that the hen is created in the image of a woman and the rooster in the man's, must have not paid attention. Nature points its finger: Pay attention! The rooster's empty, bulky strides and the hen's foolish loquacity are serious warnings. A normal man should blush when he sees a rooster. The hen, at least, can lay an egg.

In relation to its weight, the lion has the mammal's largest heart. Otherwise it would not have room for its royal disposition. Richard the Lion-hearted! A mystical spell vibrates down the spines of children. Here are history and zoology in an artistic union—both share the mythological rank and reality.

The leopard is smooth as a cat in your bed, playful and cuddly, smart and tame, content with a piece of paper on a string, but also the only beast of prey that can walk into a village and carry off an Englishman in the light of day. It is the most beautiful of all beasts of prey.

A Norwegian lynx

The great, Norwegian animal of the night, the lynx, is also an animal that can be as tame as a pussy cat, but patient and loyal, more loving than a house cat because of its intelligence. It likes to play with dogs and lets chickens alone. Only the sight of a house cat can it not abide; if these two enemies should meet, one plus one becomes one before you know it.

Then we have snakes and hummingbirds, rhinoceros, crocodiles and polar bears. There are brown bears and flounders, starfish and mussels—every animal is a word from nature.

Animals play a curious role in fairy tales. Either they are charms cast upon the prince or they are personal characteristics, not necessarily "animalistic" sides of them, but magical, secret powers that turn to the human heroes or who unexpectedly appear as helpers. In any case, the relationship between animal and human being is extremely intimate.

In legends animals appear with distinct intention; it is the "animalistic" as mirrored in more or less moral qualities. Snakes and doves, wolves and lambs are synonymous with evil and good. The lion who is befriended by the hermit in the desert and later refuses to attack him in the arena is nothing more than a clear illustration of the ideal that the hermit overcame his inner animal of prey, not killed it but tamed it, during his asceticism. The lion's wildness has become something that will help him at a later time when he is in danger. Along the way from fairy tales to legends, the animals have become less important; they have lost some of their supernatural dimension.

The picture is very different in nature sagas and myths. Here animals have retained their power and their perplexing riddles. They have nothing to do with moralistic symbolism or instructive allegories. They appear in full size, fantastic and supernatural, no longer simply animals, but more like gods that man should kneel before and pray to. Is it so strange to pray to animals? Who knows who/what they really are? The more one thinks about it, the more natural it can seem that man prayed to animal gods, as in Egypt.

The ideal animal is the unicorn, an animal that does not exist, never has and never shall—the true animal. The unicorn is the animal that has never eaten or slept beyond the limits of mythical gardens. For just a short moment it stands and stares at us through the branches of the fairy tale forest with huge, all-knowing animals eyes—then it is gone. The unicorn is so much more animal-like than other animals because it has never been corrupted by physical existence. The unicorn is itself, it is original and unique. The poet Rilke saw a unicorn in the forest one night. He wrote that from its forehead stood the single horn like a tower in the moonlight. Its coat was white.

In good zoology lessons no scientific observation should be omitted if it is important for providing a strong picture of the animal; but the other, scientific side of all animals, the unicorn, should also not be omitted. Sometimes I think zoology is a symbolic park. Who are our brothers of the night?

In our soul live unicorns, falcons, deer, polar bear, lions, rhinoceros, tigers and cows, but also the Askaladen, Aase the Goose Girl, and Lille-Kort. In animals we meet our own past. Animals are world history.

The same is true of mythology and sagas. Behind world history lie two additional world histories: zoology and mythology. They are cast forms, preserved conditions from the past. Every mythology and every animal species have that in common.

According to mythologies, human beings belong under the gods and the angels. According to zoology, human beings belong under the animals. Somehow we synthesize the lower angels with the higher mammals. That is not a bad combination. But as angels and as animals we are problematic examples. In mythology and zoology live our secret memories.

Jens Bjorneboe (1920–1976) was one of Norway's most cosmopolitan and controversial writers: a novelist, poet, playwright and essayist. His works have been translated into many languages. He was a class teacher at the RudolfSteinerskole in Oslo.

Chemistry in Grades Seven to Nine

by Jan Haakonson
translated by Ted Warren

A teacher takes on the task of conveying to young people natural phenomena as well as insights into the relationships between natural events, and preferably in such a way that they receive a powerful developmental impulse. Chemistry is relevant in the seventh grade. But in chemistry lessons a negative effect can easily take place, either a cramping effect that leads to direct antipathy for the subject or the creation of rigid and abstract mental images. The first is more often found among girls and the latter among boys. This is not chemistry's fault, for this subject can answer some of the deepest riddles in nature and in the human being.

In addition we can practice active thinking. To do so, we cannot pack chemistry into a gray mass of formulas and experiments. If the teacher starts with methods that result in abstraction, it is hard to make chemistry come alive again. The children's first meeting with chemistry must help them understand that the subject has to do with them and with the world around them. Teachers should not believe they can introduce chemistry in the seventh grade according to scientific recipes. Chemistry must be embedded in a number of subjects, whether children are learning within the world of nature, culture, art or handicrafts. The subjects should support each other, and in that way children will be engaged from many sides and can respond from different aspects of their beings.

Seventh grade: Combustion

The following is an outline of a chemistry main lesson block with fourteen lessons. It is built upon the method of providing children with phenomena they can judge themselves. We begin with combustion. We fill a large and spacious zinc vessel with all kinds of burnable materials. Everyone contributes his own items. There are many surprises in pockets and backpacks of thirteen- to fourteen-year-olds. Soon the fire blazes and the lesson takes off. We feed it with paper, wood chips, birch bark, pine needles, dried grass, and so forth. The more things to burn the better. The whole first main lesson goes to the bonfire.

In the next lesson we have a conversation about it. Yesterday's experiences are a bit more distant, and we recall them together. Before us lies a pile of gray ash, the remains of warmth, light and smoke from yesterday. It provides valuable content for discussion if we do not lose ourselves in defining what has taken place. The more unanswered questions the better.

Then we try to research on our own. Experiments with burning candles under a glass are fascinating. Everyone knows that the flame will die after a while, but few have experienced it. The larger the glass, the longer the flame can live. With this discovery we have taken a large step forward: Flame needs nutrition, it needs air.

Here is a fine opportunity to include a historical comment on fire. We can draw on a lot of mythological materials from earlier years. We start with the first time fire came to the earth, then the moment when man learned how to control it and use it. Eventually man made fireplaces that enabled him to take it inside. Smoke was a nuisance so a hole in the roof became necessary. Later we built fire into ovens and fireplaces. We learned how to regulate the air with vents. We look at how fire has become more and more distant from man. Where do we find it today in our houses if we are not lucky enough to have a fireplace? Down in a forgotten corner of the basement, built into cement and steel we have an oil burner. Through the little window we can see a flame. The rooms are heated by radiators under the window. But an evening in front of the radiator is not the same as a fireplace.

We also speak of body warmth, blushing, fever and heat waves. This awakens their consciousness. Now the children want to move on.

What happens if we reduce the air intake to a minimum? This is what we do in a charcoal-kiln and that is fun to describe. Children know the products from a kiln—charcoal, drawing charcoal and tar—but few know how they come about.

Now they learn that charcoal comes from an incomplete burning in a charcoal-kiln and that tar is left at the bottom of the kiln. Yet something else disappeared through the little hole at the top of the charcoal-kiln. What can that be?

Sulfur and phosphorus

Here we take a little detour. We burn sulfur. *Sulfurum* or *solferos* is the scientific name that means "the sun carrier." Sulfur has a close connection to volcanic areas where it may be found in beautiful yellow crystals. It burns with mystical, sluggish and blue flame that is best observed in a darkened room. It melts when it burns, and when the burning mass is poured out onto a plate, the drops fall like blue fireballs through the air. Eventually the characteristic, caustic smell spreads around the room and a coughing concert takes place among the children. Up with the windows and doors, take five minutes in fresh air!

Children remember such an experience with sulfur the rest of their lives! Then we experiment with smaller doses of sulfur, and with proper ventilation to avoid the bad smell. We capture the sulfur smoke in a tall measuring glass by letting the sulfur burn at the bottom of the glass. When the glass is filled with thick, white smoke, the spoon with sulfur is carefully removed. The collected sulfur smoke is "poured" out, a heavy gas compared with the air. We "pour" some of the gas into a glass containing water and mix it well. The gas disappears, dissolved into the water. If we taste the water, we notice an acidic taste that pricks our tongues. If we pour some blueberry juice into the water, the color changes quickly to red. The same happens with red beet juice and litmus. In this way we approach acids—sulfur's acid.

This process may be repeated and done with more powerful effect using phosphorus. As with sulfur, phosphorus smoke can be collected. It is heavy and easy to "pour" out, dissolves easily in water and colors litmus red. A new acid is "discovered," namely phosphoric acid.

With sulfur we had combustion with great warmth development and little light. With phosphorus we experience the opposite, an almost cold flame, but great light distribution. Yellow phosphorus will ignite in oxygen and must therefore be kept under water. With yellow phosphorus we must be extremely careful. It is very poisonous. Red phosphorus is neither poisonous nor self-ignitable and is more suitable for these experiments.

Carbonic acid

From these experiments a question arises, "Can we make 'charcoal acid' if we burn wood?" The students are most often convinced that it is possible. We begin by agreeing that some preparation is required in order to succeed. The type of wood we will experiment with should first have no moisture or tar in it. This is often done on a large scale at a kiln, but can also be done on a smaller scale in the classroom lab. A piece of wood is put in a large test tube sealed with a cork pierced by a glass rod. If we warm the test tube over the flame, we can watch the whole process of charcoal combustion. Soon tar collects inside the wall of the tube while a gas streams out of the glass rod. The gas is ignitable. Here we tell the children about the lack of gas during World War II and that instead of burning gas we burned "wood gas," what we see streaming out of the rod.

Our piece of wood is now charcoal, but to make it catch fire is not easy. So we introduce oxygen. We need quite a bit. For this experiment we use oxygen from a steel bottle. Among other things we notice that the burning charcoal pieces light up so brightly that we must look away to avoid being blinded. Our thoughts move to diamonds (made of similar material) and their power of light. When we add oxygen to the sluggish, blue sulfur flame, we can hardly believe the vitality of the flame. We conclude that oxygen strengthens and intensifies combustion remarkably.

With the help of oxygen we burn the charcoaled piece of wood. In a special combustion-proof glass we set the little piece of wood together with other charcoal pieces so we can view the whole process much better. Oxygen is fed into the glass while it is being warmed up carefully from the outside. As soon as the pieces of wood begin to glow, the flames are taken away and the stream of oxygen controls the combustion. The charcoal smoke is almost invisible. It is siphoned off into a glass filled with red cabbage juice. After a while it is colored red—it is a new acid, "charcoal acid," say the children. Or, if we choose, coal acid. We also observe that the charcoal gas puts out a burning light. These experiments are exciting because the gas is invisible. Now one of the children blows air through a straw down into the litmus-colored water. Shortly thereafter it is colored red, a surprising experience to many.

People breathe out acids, so there must be a combustion that goes on inside, if not an invisible flame. This topic will be revisited in the eighth grade. It is fine if some questions remain unanswered for now.

We have now burned three separate substances of very different character and with very different results. The teacher can explain that in these combustions and with the help of oxygen, oxides were created, in particular sulfuric oxide, phosphoric oxide and carbonic oxide. When the oxides are dissolved in water, we obtain the acids of the materials. All of these acids color litmus paper red.

The wonders of lime

From the fiery and colorful "world of acids," we turn to something quite different: lime and its processes. It is important in the introduction to present a rich variety of materials: all kinds of shells, bird eggs, skeleton bones and various limestone rock formations. Children should see how lime sculptures are formed. Stalactite caves should be mentioned. How did they evolve? The cycle of nature, with all of the changing formations and transformations, decomposition and edification, again and again, is a wonderful thing to tell children.

But how can we set lime in process? A lot of warmth must be used. Lime-burning ovens obtain that warmth with the help of coal, but we do not have such an oven. We must use "explosive gas flames" to obtain a comparable effect so we can observe what happens when lime combusts. That is how we introduce hydrogen gas, that unbelievably light gas, which together with oxygen forms "explosive gas flame." The fact that the flame is very warm is demonstrated by placing a steel bar into it. The steel bar quickly becomes fiery white and turns into a sparkling inferno.

This effective flame can burn a piece of limestone, marble, for a while. On the outside we see no difference and we set it aside to cool. This piece and an unburned piece of marble are the prerequisite materials for the next experiment. Two children hold each their own piece while a third pours water over both pieces. Naturally water falls right off the unburned piece but the burned one sucks in the water. Soon it becomes warm, so warm that the child must put it down. Steam rises and soon the whole stone falls apart. It becomes a pile of white powder. There is great excitement and many comments from the children. Amazing!

When exposed to heat, lime emits carbonic acid that disappears with the gases. All that remains is burned lime. If we add water to burned lime, the lime is refreshed. Slake lime colors litmus blue. Now we have introduced the bases. Bases are tricky.

We know that the carbonic acid that disappeared during the combustion was an acid. Our conclusion is that limestone contains both an acid and a base. Such connections are called salts. Now we can talk about the whole process of building walls, which had major cultural and historical significance. Burning lime has been known since ancient times. Slake lime was mixed with sand and used for masonry. When the process of hardening occurs the refreshed lime sucks in carbonic acid from the air and thereby returns to its original condition, limestone. To do so water must be emitted. Therefore new brick houses are wet and unhealthy. The hardening process can be improved by setting the bricks in ovens that release carbonic acid.

Finally we allow acids and bases to meet. We use strong salt acid and sodium lye. The combination is very strong but soon settles down. Then salt falls out, cooking salt. The acid that represents the light, fiery and airy has combined with the base that represents the heavy, earthly. Such forces are combined in salt!

Eighth grade: Sugar

The eighth grade chemistry block starts with sugar as our theme. Standing in front of an eighth grade class with a bag of sugar, I wondered how to get them excited. The tough guys demanded gunpowder and dynamite, and they were served sugar! There are many ways in chemistry to inspire teens to discover new insight in their daily routines.

First I wanted to explore how sugar relates to water. We dissolved as much sugar as possible in a little cooking water and watched the continual expansion of volume. In the end the volume doubled many times. It is amazing to see how excited the teens become when they observe simple phenomena while learning something significant about the nature of sugar. We let the sugar water cool off. First it became syrupy, then fairly tough. We hung a thin thread down in the fluid, and after a few days we had the most beautiful sugar crystals. At this point the students needed time to think about how the process can be put into action for practical purposes. They understood that this is the way to sugar glaze, syrup, jelly, marmalade, and so forth.

As the next step, we let sugar meet fire. In a test tube we carefully warmed up the sugar and noticed that it melted into a clear fluid. With continued warming it became a golden brown, caramel-smelling fluid that awakened great excitement. But we continued warming and noticed that it became darker and darker brown, before finally ending as a black, smelly mass that emitted burning gases. We continued the experiment by bringing sugar in an iron crucible with top and warming it up strongly. A black mass poured out of the crucible. Suddenly it was all over. The crucible looked comical as it stood with its top to the side, and what was left of the sugar was a porous charcoal mass. We blew a little sugar into the gas flame and observed how a little corn flamed up like a little star.

With this experiment we covered sugar's relationship to water and to warmth, and we saw that sugar is actually a product in which these polarities meet in a harmonious unity. Sugar can therefore, without transformation, be taken up in the blood of humans and animals where it contributes to body warmth. With plants it is different. Here a large part of the sugar transforms to solid substances that the plant uses to make its gestalt.

Next it was natural to speak about the creation of sugar in plants. We refreshed our seventh grade concepts of oxygen and carbonic gas and discussed how these two gases play their roles in the world of plants. A more complete description of the carbon assimilation process in plants can wait for the ninth grade. At this time it is more fruitful to work on the relationship between air, light and water. Light, air and warmth represent more the cosmic side that is above the earth while water represents the "earthly."

From the history of sugar

Honey, that represents the plant's flower-region, was known in ancient times and was considered a very important source of nutrition. During Alexander the Great's conquests from Persia to India, sugar cane was "discovered." Already a cultivated plant in India, it did not take long before sugar cane was known and grown all over Europe. But during the Middle Ages, sugar cane, which represents the plant's stem and leaf region, was considered a luxury rather than a nutritional substance. Columbus brought sugar cane to the Americas, where many plantations grew forth. This is an appropriate time to include discussions of all of the human suffering that is related to the production of sugar.

Beetroot sugar, that represents the sub-earthly part of plant life, had its breakthrough during the Napoleonic wars. When Napoleon tried to weaken England by blockade by stopping all supplies to and from the European continent, the beetroot sugar industry became a necessity. Our discussions proceed to distinguishing between the various types of sugar and the sugar-carrying plants.

During our study of glucose, we can look at Fehling's experiment, that can indicate small amounts of glucose in urine. Glucose's ability to reduce makes it possible for copper-oxide to be released from Fehling's liquids. A more blinding example of glucose's ability to reduce can be demonstrated by creating a silver mirror. Pure silver is reduced from a silver nitrate solution. Helped by glucose the reduced silver sticks to the glass walls and makes excellent silver plating.

Starch

Our starting point for learning about starch is potato flour and other types of flour. Flour can remind us of loaf sugar, but we can quickly discover the difference by rubbing them between our fingers. Potato flour, which contains most starch, is much "drier." We spread some in a glass of water. First it swims on the surface and later sinks to the bottom. But it does not dissolve. We also examine how flour reacts to fire. Rather than melt like sugar it quickly becomes charred. The flour flames up longer but not as brightly as does the sugar.

We speak about starch in plants. Sugar always retains a streaming movement through the plants. A characteristic feature of sugar is that it is found in a thinned, streaming condition in nature. The abundance of sugar that plants produce is transformed to starch. Characteristically, starch appears in numberless small grains distributed throughout the plant and remains still. The starch grains are stocked continually by the streaming sugar as a reserve. With weak light or at night, the starch grains are again transformed and dissolved into sugar. The plant also stores starch in areas that stagnate, for example in tubes and seeds that contain large quantities of starch. Trees collect starch in their trunks during the summer, and when they awaken to new life in the spring, the starch transforms to sugar that is taken into the sap streams.

Under a microscope we discover that every starch grain is formed uniquely for every plant. A professional chemist can identify the different types. The potato, which is a child of the West, has starch grains that remind us of mussel shells with eccentric middle points. The rice plant, that represents the East, has a multi-leafed starch grain centered on a middle point. Wheat is a more European product and its starch grains are formed concentrically around a middle point. These observations lead to meaningful discussions with fourteen- to fifteen-year-olds.

If we spread a little potato flour in cold water and carefully pour boiling water in it, we see that the gray-white starch grains disappear. The boiling water becomes more and more hard and streaming; the steam bubbles must fight their way to the surface. Soon we discover that we have a pudding-like mass. Every single starch grain has swelled and lost it structure in the boiling-hot water. The boundary between water and starch is washed out. We call this a colloidal state.

We demonstrated sugar with Fehling's liquid. The presence of starch can be detected in similar fashion with potassium iodide, as a dark brown-violet liquid. With even the slightest presence of starch, the liquid takes on a strong, deep blue color. Now we can experiment with all kinds of vegetables and foods. A piece of bread or potato gives a strong reaction to starch while a piece of carrot shows the characteristic blue color more spread out on the piece. The class becomes convinced that starch is an important part of our nutrition. And all of the starch-rich foods are transformed into sugar in our digestion, a process that already begins in our mouth when the foods meet our saliva. This theme can be followed up in a later block on the human body.

The starch-pudding we made by adding boiling water to potato flour has been set aside for further experimentation. The pudding is now very hard and stiff and has a strong reaction to potassium iodide. If we now add some hydrochloric acid to the pudding, an acid that is strong enough to dissolve metals, we witness a genuine transformation. The colloidal state is broken down and all that remains is a thin, streaming liquid. We let that solution cook a long while. During the cooking we take out two samples at intervals. We add potassium iodide to one sample and the Fehling's liquid to the other. What we observe is how starch gradually transforms to sugar: the longer the solution is cooked, the less reaction to potassium iodide and the more reaction to Fehling's liquid. This process is also used on an industrial scale. After removing the acid the sugar is cleaned and steamed and sold in stores as glucose, dextrose, sugar cane, and so forth.

Before we leave the starch theme, we create a starch by finely grating the potato. The grated potato mass is mixed in a beaker with water and set aside for a while. At the bottom of the glass lies a fine snow-white powder—potato starch. This is how simple it is to create potato flour industrially. We also spoke about making homemade alcohol.

Cellulose

Cellulose is also made from the plant's living sugar stream. Cellulose cultivates no grain as starch does, but chemically both are closely related. Everything that gives plants form is cellulose, from the finest nerve network in the flower and leaves to the fibrous stalk and down to the root system. The purest cellulose in nature is found in the plant's fruit hair, for example in the cotton plant or the bog grass. Flax also has a long tradition of application and is worth mentioning. The luster of a bundle of flax thread provides opportunity for many associations.

Cellulose has great resistance to chemical and mechanical stress. Therefore when all materials are removed by chemical or mechanical processes, leaving a pure cellulose, cell substance, this is a good base material for the paper and clothing industries. We took the time to describe a spruce tree's path from the forest to the piece of paper in front of us.

Cellulose is a substance that is eaten in large quantities but has no nutrition. The grazing animals are able to digest cellulose. We notice how a plant's sugar stream is such a mighty transformer. But when a plant transforms sugar to cellulose it no longer can turn it back into sugar, as is the case with starch. A permanent, stiff substance is created.

A bit about perfume

In defiance of the "hardening process," the sugar stream moves in a continuously finer and finer substance to the flower region and transforms the flower in color, pollen, and scent. We should not resist touching upon the manufacture of perfume, as part of our study.

The home of perfume is France, in particular an area known as "The Garden of France," a sun-filled garden with excellent climate and protected fields where Catherine de Medici built a garden for making perfume. In this countryside the air has been filled with flower scents since ancient times. Endless fields of roses, violets, carnations, hyacinths, narcissi and mimosas—all of these leaves to be made into perfume. The scents have since been reproduced in perfume manufacturer's laboratories to be made into clouds of scents that are sprayed all over the world.

Every month has its color. In the spring the violets spread out their beautiful blanket, followed by the Easter lilies' golden bells, and on and on! Early in the morning, while the flowers are still wet from the mist, the leaves are picked by hand. They are placed in large baskets before taken to the factories where their valuable essence oils are extracted. This can happen in three different ways: distillation, transferring the essences to fat substances, or washing out the essences with petroleum ether. Huge quantities of flowers are needed: 1000 kg orange flowers to extract 1 kg of essence, and 5000 kg of rose leaves to make 1 kg rose essence. The right distribution and mixture of the valuable essences is the great art and secret of perfume.

Albumen

Our next theme is albumen. This substance is specially related to living forces. We look first at chicken eggs and describe how in just three weeks the albumen that is inside the shell creates a new animal with all of its organs. From the albumen come many processes that we know as feathers, neb and claws. If the teacher is able to describe such processes, the following experiments will more easily guide the children into the secrets of this substance. What is special about this substance?

It is very fluid but not like water, a state between moving and hardness that gives it flexibility but keeps it from flowing away. Heating it up does not make it more pliable, but it becomes fatter, it becomes stiff; therefore it can carry so much life.

Fats and oils

Fats and oils are the next step. Once again we begin with the world of plants. Fats and oils are created in the seeds, where lie the germs for new life. The oil from sunflowers, cotton and olives are well known to everyone. But how the oil is extracted by cold pressing, warm pressing, or extraction and then used as food oils, animal food, and so forth—these are interesting topics for most students. Also in the animal kingdom among the warm-blooded sea animals, fats and oils are created on the periphery of their skins. The outer layers of fat support the inner warmth processes and protect them from the outside cold.

To understand what fats are, it is important to look at their consistency. Fat has its own form and does not crystallize in the transition from liquid to solid. Even when fat stiffens it retains flexibility in a smooth, butter-like consistency. Here warmth is at work as in no other substance; every heating or cooling raises or lowers the degree of flexibility and movement. Nowhere else in nature is warmth expressed in this way. The same is true for oils. For our study we choose experiments with practical uses in daily life: grilling with fat, oil in water, fat and oil meeting fire, and so forth.

Very characteristic for oil and fat is their relation with water—the substance where most life processes and chemical reactions take place. Fat and oil separate sharply from water by floating lightly on its surface. Even after you shake oil and water together vigorously, still the oil will separate from the water, float to the surface, and form a new, cohesive layer. To the contrary fat and oil can dissolve in liquids such as gas and ethers, liquids that ignite readily, but themselves do not mix with water. We also discussed how to make butter, margarine, candles and soap.

Ninth grade

Now the children are in the middle of puberty. They are not only physically mature, but they are slowly becoming mature in their lives on earth. If you choose to help young people along a healthy development, you must awaken interest and engagement in the world. Teens have enormous interest and take much enjoyment when they deepen their understanding of everyday things. The role of a teacher is not to give them certain opinions but to be as objective as possible when showing them where the problems lie. A teacher can support the students with opportunities to figure things out themselves. Alcohol and the arms industry in relation to Alfred Nobel are good themes to address at this age. A natural starting point for this block is:

Photosynthesis

When we study carbonic acid assimilation, we introduce a transformational process that provides a basis for plant life, animals and human beings. It is called photosynthesis. Jan Ingenhousz, the man who discovered photosynthesis, described it accurately in his treatise of 1779. His title was: "Experiments upon Vegetables discovering their great Power of purifying the common Air in the Sun-shine, and injuring it in the Shade and at Night."

> I registered that the plants do not clean the air, as [Joseph] Priestley claimed; first after six to ten days, but that they complete this valuable process within a couple of hours. The origin does not take place in the plant's growth, as Priestley stated, but in the influences from the sun rays. I found that plants have a remarkable ability to transform the air taken up by the atmosphere into truly oxygen-rich air. This cleaned air flows continuously from plants and enables the atmosphere to support life. The lighter the day, the more the plants receive sunlight and the faster the process continues. Plants that stand in deep shadows do not meet the prerequisites for cleansing the air; they give off destructive air that disturbs the atmosphere. Not all parts of the plants can clean the air, merely leaves and the green stems. Bad-smelling, poisonous plants have the same ability to clean the air as healing plants. The strongest, oxygen-rich air streams from the underside of leaves. All plants pollute the air at night.

This theme provides an excellent educational moment, a concrete starting point for looking at a mysterious process that the children can understand. We are at the core of pollution and ecology. We realize that pollution cannot be stopped easily or quickly but must be addressed through a new way of thinking.

We demonstrate how photosynthesis and breathing take place and how the gases relate to each other. If you breath into a glass with a limewater solution, we can prove carbonic acid in our exhaling as the limewater becomes paler. We revisit our experience of carbonic acid in the seventh grade.

We demonstrate the plant's ability to develop oxygen in sunlight by using water plants. The oxygen that is developed underwater is collected in a funnel with a rubber hose and a clamp. The collected oxygen is proven with a glowing wood chip. We also revisit our experience of oxygen from the seventh grade.

It is useful to draw a plant that stands between heaven and earth, surrounded by air. Water is taken in by the root with a certain amount of salts that stream through the stalks and into the leaves. In the heavenly part of the plant, carbonic acid is taken in from sun forces. There are also starches and sugar in the leaves. Now is a good time to present a simple formula. Photosynthesis can be formulated as: Carbonic acid plus water gives sugar and oxygen.

A glowing wood chip is held over the glass filled with limewater. It fades. Therefore: Carbon and oxygen become carbonic acid. But in the plants the opposite takes place: carbonic acid minus oxygen gives carbon. Yet we find nothing black in a plant, no hardened carbon, but rather starch grains and dissolved sugar that are flexible, moving, life-giving substances. Sugar is created by carbon and water; we call it carbon hydrate. It is living and moving carbon. Carbon plus water gives carbon hydrate. In other words, air is the plant's coal mine.

We should also take a look at sugar combustion within the human being. It is the opposite of photosynthesis. We take in sugar and oxygen and breath out carbonic acid and water. Sugar plus water gives carbonic acid and water. Sugar and starch were already characterized in the eighth grade, and now we can approach these substances more scientifically. From this starting point we can visit many new areas of chemistry.

Alcohol

To demonstrate the fermentation process from sugar, we begin by making a "composition" that is then set aside. In the meantime we make experiments to see how the mixture emits carbonic acid during fermentation and a special smell appears. How do you separate the alcohol? With heat and an ingenious apparatus, we can separate out different substances from a mixture by converting each liquid into steam (gas form) and then cool and condense it to liquid again. This is distillation. I explain how we can make alcohol starting with starch from potatoes or grain. The starch must first be changed to sugar, as we learned in the eighth grade.

Wine grapes provide an alternate starting point for making alcohol. First the grapes should be described in detail. We are told that if you cut a grape vine at the root, a waterspout can be sent thirty yards in the air from the pressure the plant creates at the root. When you explain how much warmth the plant demands, that the farther south it grows the sweeter it becomes, the students understand the polarity that is united in these grapes: fire and water. If you let grape juice ferment, carbonic acid will be emitted and the sweet taste gives way to a strong, burning taste. Sugar becomes alcohol and carbonic acid.

Here is short description of how alcohol works in the human body. Sugar is taken directly into the bloodstream; only the liver can regulate the percent of sugar. Alcohol skips past the liver so that the alcohol spreads throughout the entire body. The basic sugar in the body that allows for harmonious body control and willpower is replaced with alcohol and a false feeling of body control without our abilities being improved. Instead of a peaceful warming effect, a condition for exaggeration takes place. One becomes weak in personality and self.

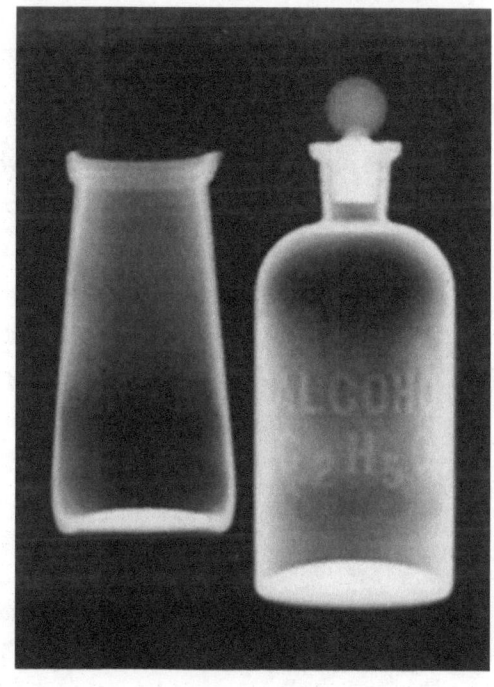

In the classroom we can experiment with alcohol. Its mobility and ignitability plus the unlimited dissolvability in water are characteristic.

Sometimes pupils prepare arguments for and against the use of alcohol. As they are immune to moralizing, it is often more effective to approach the issues of alcohol abuse from the insights they have gained in chemistry. What does the abuse mean for the individuals, families and society as a whole? Can the problems be solved with stricter laws or higher costs for alcohol consumption?

Our experiments can take their understanding of alcohol to another level. A fermenting mixture of alcohol with potassium oxide produces ether. The ether has both increased mobility and ignitability, but the ability to dissolve is reduced. We know how ether works on people from ether narcosis. While the self is weakened under the influence of alcohol, it is shut down completely by ether narcosis.

It is important to record the alcohol tables in order to characterize the various types. Other tables such as plant acids, fat acids and their salts can be written down and memorized.

The chemical industry: Alfred Nobel

Our study of fats and oils is a continuation of the eighth grade study. In the ninth grade we approach industrial production and uses of margarine and soap.

We can obtain fat acids and glycerin when we separate fats. Glycerin leads us into the study of explosives. Alfred Nobel's discoveries and adventurous biography are valuable resources. We describe his greatness, all-sidedness and noble character and also his deeply problematic endeavors. He was split between mind and heart. The work he did with glycerin paved the way for new weapons while his heart protested against the use of weapons. From him we learn about the concept that remains in our modern times: Prevent war by developing weapons! The revolutionary character of dynamite as used in great tunnels casts a dark shadow over the Nobel Peace Prize when you realize that the prize money comes from the weapons industries.

By treating fatty acids and their salts, we learn about stearine acids and their salts that produce light. Stearine candles have a long history and their development is worth describing. We burn our candles and enjoy the atmosphere but few realize how much work went into developing the easy-burning, drip-free object. Notice how the wick bends away from the flame's area. Therefore it burns out and disappears on its own. This was designed a hundred years ago by weaving together three threads but allowing one thread to be shorter than the others. The teacher can refer to Michael Faraday's book, *Lectures on the Chemical History of a Candle,* a series of lectures given at Christmastime in 1860 to a group of young boys and girls. Faraday teaches us how to observe.

While we are learning about fat and oil, we also discuss crude oil. How the oil is brought out of the sea is a theme that excites ninth graders. It is processed at huge refineries, where distillation technology has reached it high mark, a technical triumph. In the tall, distillation towers, crude oil's various components are funneled to different zones according to their boiling points. We are familiar with these products from everyday life: gas, petroleum, heating oil, lubrications, Vaseline (petroleum jelly) and asphalt. We are dependent on all of these products, and ninth grade is an appropriate time to talk about responsible use.

We have seen a few moments from chemistry lessons in the seventh, eighth and ninth grades. We can conclude that lessons are most successful when they relate to the themes and challenges with which the students are familiar. Because chemistry is a large part of everyone's daily life, our main lessons can quickly move in unexpected directions.

Astronomy: The Oft-Forgotten School Subject

by Sven Bohn
translated by Ted Warren

Astronomy has never enjoyed a distinguished place among modern school subjects. Despite the decisive meaning of our globe's daily rhythm and the changing of the seasons for all life on earth, the study of astronomy lacks a certain relevance that would bring more focus to it as a school subject. A few sections can be found in the final pages of geography books, a situation that not even modern space travel has been able to change.

Is this exaggerated? Do most people know the modern scientific worldview of our solar system with the planets circling in continually wider elliptic paths? And that the shining stars are newer suns in space, perhaps with planets circling? And who does not know there are star clouds that in truth are huge Milky Way-like system galaxies far, far away? Most significantly we associate astronomy to the foundation of reliability: magnificent [astronomical!] numbers. Temperatures of millions of degrees, massive formations as large as entire planetary systems, densities thousands of times more concentrated than gold. And most impressive are the enormous distances: Just to the moon is as long as circumnavigating the earth. It is 150 million kilometers to the sun, and the closest fixed star is four light-years away. Does everyone know what a light-year is? The diameter of the Milky Way is 100,000 light-years, enough to give you goosebumps when you read these numbers. Kant must have felt this when he spoke of devotion to the starry heavens above and the moral laws within. Did someone say education in our times is weak in astronomy?

Very few people know the starry heavens. In general, beyond the Big Dipper and perhaps Orion's Belt, people know few other constellations. They cannot tell the difference between a star and a planet, do not follow the changing phases of the moon, do not know how many degrees the sun stands above the horizon, do not even notice the changing positions of the starry heavens when they are on vacation in Greece. The simple fact is that the stellar sky is becoming more and more closed for mankind in pace with increased access to general knowledge. Some have an intellectual relationship to astronomy, others continue to walk outside and recite their favorite poet, for example Henrik Wergeland:

> How Venus shines tonight.
> Do the heavens also have a spring?

Poetry is only powerful when our feet are not frozen! Therefore few successfully overcome the schism we encounter here, and such a human schism it is: between experience and knowledge. It appears to be the signature of our culture that this schism will not be closed but remain an open wound within each person.

Is it possible to close it? Rather than further theorizing, I simply ask, "How can we achieve a relationship to the starry heavens we observe? What we see with our eyes matches up so poorly with what we know. For where in the world is the solar system? What is so simply and delicately laid out in textbooks is impossible to see, either by day or by night. It is simply not possible to discern between near and far, everything is chaotic and not even the Big Dipper that was behind my neighbor's house three months ago is still in the same place. To my surprise nothing I have learned in books can be used; observation and concept collide head-on.

What we have learned in astronomy are models of thought won over centuries and millennia of battles for knowledge, by geniuses in crisis and hard work, in desperation and ecstasy. For us, if we have not taken part, all so briefly, in how it was acquired, it is simply dried-out knowledge, not worth the paper it is written on. In astronomy the models actually stand directly in the way of our observations. If astronomy shall have a place in school that can help children create a more wholesome relationship to our cosmos, we need to begin at the other end and ask: What do we see?

The first main lesson

What do we see? With that question we begin our lessons in the astronomy block in the winter of the seventh grade in Oslo, Norway. We start with the drama that takes place outside, the archetype of our experience of the world: the sunrise. The pole of life, light and darkness meeting in the long winter night, is briefly relieved by the short winter day and we can study the movements of the sun through the sky. How does it truly move? And how was the sun a week ago? How was it at New Year's Eve? And what will we see in late spring? We all know a lot, the shortest days are near the new year, the longest at the end of June, and there is always someone with a peculiar explanation: Everything is due to the slanting of the earth's axis, something she read in a book.

It is important not to go too fast. First we shall get to know the sun's daily movements across the heavens. It rises at a slant, climbing gradually slower before it reaches the top of the arc at midday. Thereafter the sun follows a symmetrical movement down to the horizon and then it sets. When the daylight is shortest, the arc is smallest: It is a low winter sun. Yet the sunrise slowly changes position, moving more toward the east, and accordingly it sets more in the west, while the middle point becomes higher overhead. The midpoint is always in the south. Yes, that is the south! And if we know the south, we can find the other three directions. Then we ask: How far into the spring do we go before the sun stands up directly in the east? The vernal equinox, March 21, and the corresponding fall equinox, September 23. On those days the sun rises in the east, sets in the west, and day and night are of equal length.

Homework: How many days does the summer half-year have and how many does the winter half-year have? The surprising answer is: 186 for summer and 179 for winter. The explanation—we wait a bit!—What do you think? If someone answers: The earth moves faster in the winter, that is correct, but, again, this is something she has observed and not just read. As a teacher I can imagine answering, "Next year you will learn about Johannes Kepler who discovered the law behind these phenomena." But first we need to get to know the starry heavens.

To do so we need to know the sun's daily movements. We must know the movements well enough to imitate them; we stand up, turn to the south and point with our hands. Many think this exercise is much too simple and consider it superficial. But they are fooling themselves, for these movements are the most important to understand. They are the same movements all heavenly bodies take during the day, as observed from the earth. During the first days we work with this pattern that is taken daily by all heavenly bodies (except for meteors and satellites that are influenced by the earth's rotation).

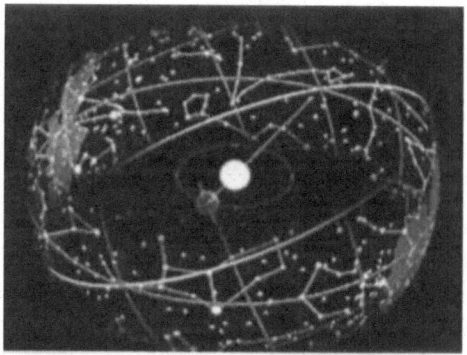

The Earth in its orbit around the Sun causes the Sun to appear on the celestial sphere moving over the ecliptic, which is tilted on the Equator.

Our calendar makes for an interesting study. Right away there are problems involved. The sun's yearly movements must be presented, a movement that cannot be observed directly because of the sunlight, though it has been known for centuries. The sun's movement in the Ecliptic through the twelve signs of the Zodiac has been known since the Babylonian civilization, an impressive achievement. I always challenge my pupils to figure out how it was possible to discover. How long does it take the sun to move once around the Zodiac? We all know—a little less than 365.25 days. So how can we best assess the extra one-fourth day? This is a piece of practical mathematics that has been resolved in a variety of ways through the ages, with each way of solving the fraction a reflection of the cultural epoch that created the method. Think, for example, about the Egyptians who did not have a Leap Year and therefore experienced a slowly backward shifting of the seasons, such that after 1440 years this amounted to one full year. The "mistake" reflects the slow rhythm that played an important role in the Egyptian culture.

The next step is to work with the many phenomena of the moon. Naturally the most well-known phases are the full moon and the new moon. But in which direction does the new moon open? At what time of day do we see it? With effort we order our observations: The new moon is an evening phenomenon and it opens to the left, away from the sunset. The new moon waxes from evening to evening, while it becomes visible for longer periods of time. And when it is full, it stands opposite the sun. As the moon only reflects sunlight, its dependence on the sun is obvious. Its cycle is reflected in its name: one month. Twelve times each year the moon grows into the full moon, hence our twelve months.

Invariably someone will ask: Does it happen accurately? No, there is an error of roughly one day per month, which adds up to twelve days per year. We can speak of a "moon year" as 354 days, something that plays a role in many cultures, for example in Islam, where the month of fasting, Ramadan, is calculated according to the phases of the moon and therefore moves backward each year by twelve days.

The course of the moon is not so complicated. It is interesting to study how the course changes from month to month and how it happens characteristically within the seasons. The first thing we notice is that the new moon in the spring is very easy to see. If we consider the slanting of the sun's path, it is easy to understand why. For the most part the moon follows the sun's path, the Ecliptic. From February to March the sun climbs many degrees higher in the sky. The moon accomplishes this same climb in just two to three days. Therefore it becomes visible high above the sunset. The spring new moon is a well-known sight. It eventually reaches the sun in the signs of the Zodiac during the summer, but it moves quickly toward the signs in the fall. Therefore the full moon in the spring is not especially high in the sky. To the contrary, at the new year the full moon is high because it stands opposite the low winter sun and thus reaches its fully-lighted phase in the summer constellations, Gemini and Cancer.

During the summer it is the opposite. The sun stands high in the sky while the full moon, now in the winter constellations of the Zodiac, hangs low on the horizon just above the islands on the lakes. The waning moon is a less considered phenomenon. It is observed by roaming poets who come home early in the morning and write about how beautiful the new moon shone above the treetops!

We must wait until the early fall mornings to observe the waning moon while it follows us on the way to work. We all remember the special feeling of the waning moon in October and November; high in the east it lies with the opening to the right and the morning sun "behind" it toward the horizon. In class, whenever I speak about the waning moon, the students have a whole reservoir of experiences they bring forth. And when they go home after classes, they see the same moon they have seen hundreds of times as if for the first time!

That is the point. Daily observations are the most fun. Children can be filled with impressions from sensational space explorations or telescopes, but it is better to know what our own eyes can tell us. The sun and the moon, the calendar, the Zodiac, and an evening or two under the stars—this is already a lot! We should let it rest for some months before the next astronomy block. Or maybe we can fit in an orientation among the heaven's other constellations or draw the globe of the heavens in which we stand, while it moves once each day? This is what we need to become a stargazer: a truly naive, geocentric model of the heavens, in which all of the bodies are attached to the inside of a globe where our sight creates an unlimited radius from the center of the globe. This model shall be our helper for a long time forward.

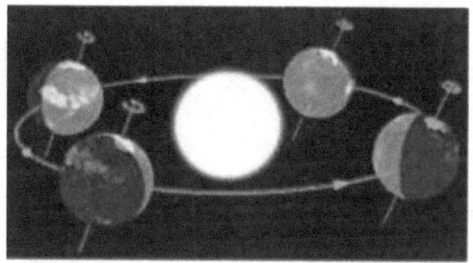

Diagram of the Earth's seasons as seen from the north. Far right: December Solstice

At this point we can weave the important threads into a geography study and work with the children's understanding of the earth's position in space. If we travel north to the heaven's North Pole, which in Oslo is sixty degrees over the horizon, we increase degrees in the sky by one degree for each line of longitude we pass. At the same time the heaven's equator will sink accordingly, and eventually the sky of the southern hemisphere will disappear below the horizon. If we already know the path of the sun, the phenomenon of the midnight sun is easy to understand. Yes, also the moon and other heavenly bodies become visible all day long! But let us continue our thoughts concerning the polar regions. What happens there?

Only the northern hemisphere is visible, but it is visible day and night! And what is the result? Of course we have the midnight sun for half a year, from the spring Vernal Equinox to the fall Vernal Equinox. For the other half year the sun goes down and we have a half-year of winter night. We can also say that the rhythm of a day and the rhythm of a year fall together. The year becomes a day.

We will experience just the opposite if we travel south. The heaven's North Pole sinks while the heavenly equator rises. Eventually the North Pole lies on the horizon in the north and at the same time the South Pole becomes visible in the south. We are on the Equator: Within a day the whole starry heaven passes over us and the heaven's equator lies right above us on the zenith, and right in the east and the west. The day is almost like the whole year, divided in two by day and night, and the seasons are gone.

A thought for reflection: Between these polarities we humans live our lives: the southerners more strongly in the daily changes, the northerners more strongly in the change of seasons. Where are these elements found in balance?

The second main lesson

A less accessible area for children is the study of the planets, their rhythms and movements. But it is within their reach! In the eighth grade it is natural to study the planets, as it meets their need to observe something that requires patience. For many fifteen-year-olds this can be a balance to their restlessness and unfocused self-indulgence.

Before we start this block, it is wise to work with the starry sky above, something that can be made more accessible through the Greek sagas that provide the background for the names of the constellations. Perseus, Cepheus, Cassiopeia and Andromeda are some of the more important constellations, but also the Hunter Orion with the Dogs, the Great Bear, the Little Bear, and Hercules are some of the easiest to learn. I begin with the last two stars of the Great Bear, also known as the Big Dipper. These give us a starting point for the circumpolar constellations (those that never go below the horizon at our longitude), and then we can move out to the larger oval that is created by the brightest stars: Capella in Auriga, the twins Castor and Pollux, Procyon in the Little Dog and Sirius in the Big Dog, Rigel (one of Orion's feet) and Aldebaran, the Bull's red eye, not far from the well known seven stars of the Pleiades. The entire oval is a wonderful section of the heavens that is entirely visible on clear winter evenings in the south.

Then we can approach the Zodiac again, now from its various positions. It runs through the above-mentioned oval in the sky and when it is in the south the Zodiac has its "high" position in the firmament, high because we see the summer constellations. But also the Ecliptic's two slanted positions are important; the eastern slant is filled with the evening constellations of the fall and the western slant is typical for the March-April evenings. The Ecliptic's low position on our longitude with light summer evenings is not an unknown sight. But it is more effective to observe that part of the Zodiac at another time of day, preferably when one of the planets is there. The constellations in this area—Scorpio, Sagittarius, Capricorn—are relatively inconspicuous constellations and we may need a planet to help locate them. This may be a good way to begin a systematic approach to planetary observations.

The five visible planets can be placed naturally into two groups: Venus/Mercury and Mars/Jupiter/Saturn. We will not yet use the descriptions "inner planets" and "outer planets." First we need to observe the phenomena that make these descriptions natural.

Let's begin with Venus. The queen of the heavens has amazed and surprised many people for eons. In light intensity it can outshine all heavenly bodies but the sun and moon. Many people do not realize that it can be seen in full daylight. Like Wergeland, many have noticed Venus on clear spring evenings. An intense white light shines above the sunset before any other stars appear. And one can observe it week after week, month after month. It is easy to see that it is fairly "close" to the sun, and increases the angle slowly until, after a few months, it reaches an approximately 48 degree angle to the sun and sets three to four hours later each day. Then it remains steady before getting "closer" to the sun again.

A short while thereafter the largest angle follows the maximum light strength. We can assume that the reason for this is that it is closer than before but not so close to the sun that the sunlight limits its clarity. In this phase it can be seen at midday, though it does not last long. After a few weeks in the sunlight, it disappears. If we were to make a drawing of its path, a beautiful orbit would appear including a meeting with the sun in the middle. Soon it reappears from the "other" side, as the morning star. Relatively quickly it reaches the same maximal degree, turns around and is pulled fairly slowly toward the sun again where it reaches its western light maximum and disappears after a half year in the sunlight. This time it is not visible for a longer period of time, almost half a year, and then it finally reappears to become, once again, the evening star. It has completed a synodic cycle.

Many viewers are disappointed when in the following spring they do not find Venus in the same place as the previous year, but that is not the planet's rhythm. The cycle around the sun, as seen from the Earth, takes 584 days or roughly 19 months. What is the Venus rhythm in relation to our year? Five Venus cycles are almost identical with eight Earth years. Only two days make the difference. The Venus cycle lasts almost exactly three-fifths of a year. Practically the same Venus cycle we witnessed in 1983 was repeated in 1991, 1999 and 2007. An eight-year rhythm is apparent, a period of time during which Venus traces a pentagram, a five-pointed star in the Ecliptic with its orbits, harmoniously drawn with 72 degrees space between. If we draw a diagram of Venus for the eight-year period we will know where to look for the planet at any time.

Our accessibility to Mercury is nowhere near as easy. The moving war god is volatile and demands a lot from those who search for him. It is told that Copernicus never experienced the planet with his own eyes, something not only caused by the meteorological conditions at the Baltic Sea, but also because he was not a practical astronomer. Instead, his contribution was to solve the riddle of the planetary orbits, something we can do best on a piece of paper!

In principle Mercury moves just as Venus does, but its furthest degree of distance to the sun is merely 28 degrees, and it is therefore never longer from the light of the sun than a maximum of a few hours. Therefore we will find it a few hours after sunset or before sunrise and, because of the Ecliptic's angled position, only on spring evenings or fall mornings. Then it stands in a higher positioned constellation than the sun and receives something additional that corresponds to the new moon in the spring and the new moon in the fall. But we can see it for only a short period of a couple of weeks, and during such a period, we must pay attention to spring afternoons when we have clear western horizons or good eastern horizons on fall mornings. To sight Mercury requires interest and focus. Mercury has such a short cycle that you have the chance once every spring and fall. 116 days are used in a synodic cycle or just less than four months. During that time it accomplishes a cycle that covers its upper position or conjunction with the sun "behind," the largest eastern angle (evening star), the lowest position ("before") and the largest western elongation where it appears as the morning star. Then it returns to its usual conjunction.

The rhythms of Mercury are well established within one year. Three cycles is 3 x 116 = 348 days, something that makes Mercury's period of visibility for every year advance (365 minus 348, or roughly 17–18 days) "earlier." For example if the largest eastern elongation happens on April 18, the following year it will happen around April 1, then March 13, thereafter February 24, then the beginning of February and thereafter January. As it approaches Christmas and New Year's Day, we no longer have any advantage from the Ecliptic's slanted position. If we look for the next time the planet has the largest eastern angle, it falls not on April 18, as it did seven years ago, but on April 25, one week extended. Mercury's rhythm of seven years does not quite fit. The fact that this "fit" will be corrected within 46 years, or 145 Mercury cycles, can be easily calculated. The surprising fact is that this was known already in ancient times. A diagram of this planet's movement in relation to the sun will be roughly the same after 46 years.

Mercury and Venus make up one group of planets. Though their rhythms are very different, they have significant factors in common: They are both near the sun as seen from the Earth. If you understand one of them, you can easily understand the other.

We find a very different picture when we work with the other group of visible planets: Mars, Jupiter and Saturn. These planets do not move within limited angles, but regularly achieve the largest possible angle of 180 degrees. In astronomy we call this *opposition*. The planets stand directly south at midnight. At the same time they stand in the middle of an orbit that was very difficult to explain for many years. They also have their maximum brightness. The orbits, that for the inner planets take place before the sun and are therefore maximally removed from direct observation, take place opposite the sun, and even the amateur astronomer, by paying attention for several months, can follow the orbits with the naked eye. Yet each planet has its own tempo and rhythm.

It is easiest to begin with Jupiter and Saturn; both move slowly across the starry heavens. Our sun takes one year to move once through the Zodiac, the moon takes 27 days. Jupiter takes 12 years! And Saturn, 30 years, considered one human lifetime. No wonder that throughout history Saturn has been related to life and death. In the old geocentric model Saturn was the guardian of the border between the changeable and the constant, between time and eternity. Its symbol is a simplified human figure with a scythe.

Because they both move in the same direction as the sun, they are overtaken almost every year, Saturn most often, actually every 378 days, Jupiter every 399 days or after 13 months. Jupiter moves within one sign of the Zodiac for every time it is overtaken. Saturn remains for two to three years in each of the Zodiac's signs. A calculation that is fun to set up in class is an equation to find out how often Jupiter "overtakes" Saturn. It happens every twenty years. The last time was 2001.

Many conditions make Mars much more irregular and therefore harder to understand than the other planets. Why not wait to learn about it and then also prepare the transition to another theme, a historical perspective of our worldviews? It was the warlike planet Mars that Johannes Kepler wrestled with to discover the planet's exact laws of movement, later known as "Kepler's Three Laws."

In principle, Mars acts like Saturn and Jupiter. But the cycle time along the Ecliptic is barely two years and the sun will therefore overtake it approximately every 26 months. Notice the word *approximately*. The synodic cycle can vary up to 50 days and it appears to resist every regularity. Roughly stated we can say that Mars makes an orbit (in opposition to the sun) less than every two years and seven orbits within 15 years. Then it makes a new series of orbits, somewhat altered from the first. During a fifteen-year series the light intensity set its own rhythm: The orbit in 1969 showed very intense light as Mars shone at the same time as Jupiter—but red! It shone weaker for a while and was at a minimum in 1977. In 1980 it was strong again and in 1984 it was very strong! The light-intense orbits take place in a part of the Zodiac not easily observable, the fall/winter sign we know as Scorpio. In Norway, Scorpio is very low on the horizon, in New York it is fine, and in Australia it is a beautiful sight, though the sun is in the north during the afternoon!

Let us return to the irregularities of Mars. While the planet might irritate us, for a man like Kepler this was a challenge. He quickly noticed that Mars' movements contraindicated Copernicus's circle theory. Here was one planet at least that did not follow the idea of a circular orbit. The circle movement was in no way Copernicus's original idea. It was a thousand years old—the Earth as God's middle point, and everything orbiting in perfectly formed circles because they have one middle point, The One. Yet, the Copernican representation of the heavens was known even in Ancient Greece. It was reawakened but not fully accepted until 60 years after his death.

Is what appears on Earth so enormous that even a genius is not enough? Kepler is a symbol for the power of thought, the original and creative concept, unafraid and persistent. While his representation of the heavens contributed much to modern mathematics and physics, it was not enough. What was missing is the new world-historical quality of knowledge: It shall be free of all speculation; it shall be built upon precise data, by correctly collected data and ordered observation, possible errors shall be removed, and everything shall be controlled and open. Is it possible for him to find new numbers and new observations or must he rely on Hipparchus' tables from ancient times?

Yes, in Denmark a certain nobleman with a nose made of silver sat on his island. On Hven in Øresund, he and his many assistants worked through the clear, starry nights in his observatory, Uranienborg, observing, measuring and writing. Page up and page down, book after book were filled with numbers as dry as a modern telephone book. But Tycho Brahe did not share his research. It was his private property and reserved for formulating his own representation of the heavens. His numbers were not accessible, while far away in Germany, a poor boy sat near his mother's arm watching a nova light up in the starry heavens. He was 25 years younger than Brahe, from another class and nationality. If our representation and knowledge of the universe was to be developed any further, they must meet. Their two dramatic biographies take them where they do not want to go, but must go. Indeed, they met at Kaiser Rudolf's Court in Prague just before Brahe died. The protocols, filled with Brahe's observations of Mars, came into the hands of the only one who could put them to use—Kepler.

A restless search for the truth began. A heavy battle with numbers was fought with new mathematical models, doubt and anxiety. For Kepler, if anyone, was a deeply religious man. In him the decisive thought arises: the ellipse! But it has two focal points! Yes, but it must be an ellipse: One focal point is empty, in the other stands the sun.

Mars and its strongly eccentric orbit caused great interest in the planetary orbits, and clarity for a large part of our scientific and technological development. There are many reasons to work thoroughly with astronomy. The portal to the new science we praise so highly is Kepler's work with the orbits of Mars. By deepening our understanding of the biographies of Brahe and Kepler, the students' respect for both poles of human knowledge can be awakened: observation and thinking. Both are necessary parts of a whole, the very same whole we strive for in astronomy lessons at the Waldorf school by allowing the phenomena to live and grow into science.

The Starry Heavens and Our Self

by Jørgen Smit
translated by Ted Warren

Modern astronomy has provided an incredible amount of knowledge about the stellar heavens. With the help of enormous observatories and powerful computers, all of what we know about the phenomena of the heavens is now charted with more precision than ever before, our observations double what was previously available. We can participate in this knowledge through large encyclopedias, thick astronomical books and popular science accounts. But has all this development and accessibility of information resulted in a closer, human relationship to the stars than we have had before? Or has it, to the contrary, actually distanced us from a more intimate experience of the stars?

Modern astronomy presents a model of the starry heavens that resembles a huge machine. And despite the fact that many propose there is probably life and consciousness somewhere else in the cosmos than on earth, it is still unknown. We are most interested in the purely quantitative content, the cold, hard facts, for this kind of information can be manipulated by computers. The notion of life and conscious beings on other planets is only used to balance out the known facts in all kinds of fantasies in comic books and novels, just as distant from reality as the knowledge that we now have concerning the starry heavens.

And what about the "knowledge" most adults have today concerning the heavens? Of course everyone "knows" that the earth and the planets move in elliptic paths around the sun. Everyone "knows" that the stars are suns that are extremely far away, and so forth. But who "knows" the stars when they appear in the heavens at night? In which constellation does Jupiter stand tonight? Point it out! In which constellation did it stand a year ago? Point it out! And in which constellation will it stand in one year? Point it out! To such specific questions one cannot expect an answer from anyone but a professional astronomer or from those few people who love the stars so much they follow the starry movements over the years.

But this is exactly what one needs to do if one is to truly get to know the stars. Learn the constellations by heart and understand how the planets move so one can point accurately around the starry heavens. But what has one accomplished with that? At best a small beginning. If nothing more happens, it will all be forgotten, as quickly as other superficial knowledge. It would be as if one learned by heart all of the names of people at a party and could point them out perfectly: that is Hansen, that is Petersen, and so forth. If one does not get to know them better, their names will quickly be forgotten. One can only get to know people by living with them for many years. It is the same with the stars. They can become our friends; their names become secondary. The qualities of the starry heavens can be known. But to do that we need something for which our hectic tempo and superficial lifestyles do not give much room.

We need quiet observation in which we give ourselves time to linger with every single thing so that the important aspects emerge from the unimportant. If we live with the starry heavens for many years, we discover three main areas, three spheres. More correctly, we rediscover them, for they are obvious, simple observations that everyone already knows, that everyone has seen, but which we do not notice because we race right on to other daily activities or complicated astronomical calculations. When we have lived with such simple observations over a longer period of time, we will notice the depth of them, they are rediscovered. Here they are:

1. If we look into the northern heavens on a clear, starry night, we notice one of the few constellations we know: the Big Dipper or the Great Bear. This starry figure stands in different positions during the night and during the year. Yet the figure's form is the same as when we were born, unchanged during our lifetime. So have these stars lit up the skies for our great grandparents and for people in the Middle Ages. In the same constellation they lit up the night skies for the Egyptian pharaohs five thousand years ago. Really? Have they not changed position even a tiny bit? Yes, they have, but only a tiny bit. The changes that have been calculated are so little that two ten-thousand year periods must pass before the constellation will become something very different.

Stars twinkle because ... they're so far away from Earth that, even through large telescopes, they appear only as pinpoints. And it's easy for Earth's atmosphere to disturb the pinpoint light of a star.

2. It is quite different with the planets. Planets shine more steadily because ... they're closer to Earth and so appear not as pinpoints, but as tiny disks in our sky. And they change position in relation to the fixed stars from year to year, from month to month, day to day. They travel in rhythmic periods and every planet has its own rhythm. For example, follow Jupiter's movements for twelve years so that it has moved around the whole fixed starry heaven; then it is no longer merely a theoretical number—12. Every year Jupiter moves in a small arc through one of the twelve signs of the zodiac. As I write today it shines in the constellation Taurus, as it did also in 1942 in the middle of World War II, and as it did in 1930 before Hitler came to power. So there are two Jupiter periods, 1930–1942 and 1942–1954. How different is the Jupiter quality from the fixed stars' quality? Both project beyond the day's shifting mood. The fixed stars project all the way into the unlimited, eternal sphere. In contrast the light of Jupiter relates the moment's unregulated diversity with eternal peace in a clarifying, steady, majestic rhythm of time—12-year rhythms.

3. A star is falling. In a bright arch it shines for a moment and is gone. A shooting star. This brings us to the direct opposite of fixed stars and planets: the unpredictable and chaotic world of the moment. But are there also certain laws that pertain to meteors? They usually arrive in certain directions at certain times. That is so. Yet at the same time they are merely a collection of individual phenomena that are not calculated in advance. When we see a person in deep sleep, we know it is unlikely that he will suddenly begin drumming his fingers. If I see a person I know in a situation where he usually drums his fingers on the table, then it is more likely that he will drum his fingers. But it is also possible that this does not happen this one time. When we describe the qualities of the meteors (it pertains to most atmospheric phenomena), we approach something which, in relation to its irregularity, can best be compared with human body movements.

When we look up at the starry heavens and live with them over a longer period of time, we can observe:

1. The light of fixed stars in never-ending, "eternal" peace with "unmovable," solid structural form.
2. The planets and also the sun (in relation to the earth) and the moon in rhythmic movements that exhibit the moment's changeableness and "eternal" peacefulness.
3. The atmosphere's meteors that unfold in the moment's changeableness.

These three qualities can be observed in a similar manner in the human body:

1. The basic structure of the body, the formation of the skeleton that is fairly static for a long period of time, just as the structure of the fixed star constellations. The fact that mankind has been aware of these relationships is evident from numerous drawings of the human body with the signs of the zodiac placed in specific areas.
2. The life processes—blood circulation and breathing, and so forth—in which there is constant motion as rhythmical and harmonious as the movements of the planets, the sun in relation to the earth, and the moon.
3. The limbs' changeable movements that happen from one moment to the next, without particular rhythm like the shooting stars and the atmosphere's shifting life.

When we look up at the stars, we see an endlessly large "human being." We see our Self in a bigger picture. In the gigantic cosmos we find the same order as we find in our own body. The same creative forces that live in our body have formed the universe by the same laws through millions of years. If one becomes aware of these relationships, this picture can be enlivened and made more complete, step by step through personal experiences and observations or with observations shared by others. The value is found in the fact that we become neither one-sided in a "spiritual" sense nor one-sided from a "material" perspective. The material is observed as it is formed by the spiritual, and the spiritual is not something hypothetically assumed, or beyond the world of the dead, or in an inaccessible transcendent-metaphysical world, but rather something here and now.

The human being is no longer split from the world of nature, no longer unknown in an enormous, dead machine. What lives in the human being is expressed in nature because it is all the same forces that created nature. Ancient myths and original fairy tales are often created from the same perspective. The same is true of a number of ceremonial symbols. Let us take one example: The butterfly that emerges from the green caterpillar was often used in ancient times as a picture of the human being that left its body after death. But the same power that eventually appears in the butterfly's colorful life initially works hidden within the green caterpillar. The green caterpillar must die and it does die in the cocoon state. The caterpillar no longer exists as caterpillar. But from the dead, the hidden force appears in a totally new way.

The same is true of the human being. The soul-spiritual nature lies within the human body as in an empty case. It works in the body during life on earth. But the earthly existence is merely one form. Human beings die and do not continue to live after death in a thinned-out "spiritual" copy of the past, nor does the butterfly remain a thinned-out copy of the caterpillar. Life after death for the human soul-spiritual being is just as different from life on earth as the butterfly is different from the caterpillar.

This thousand-year-old ceremonial symbol is not used merely as a beautiful picture. It is used in relation to the insight that the same reality that appears in human existence at death has worked creatively in nature and therefore the butterfly's "resurrection" appears from the dead cocoon.

When children reach the age of seven, they begin to free themselves from their surrounding environment. An independent soul life begins to emerge slowly. In their ninth or tenth year, their independence enters a more confrontational phase with the outer world: my Self and others, my Self and adults, my Self and the world around me! At this time children begin to question in new ways in order to understand what life and death are all about and what happens with them as human beings after death. They ask for something higher, whether or not they put it into those words. They may pick up adult words on the subject and therefore cannot comprehend or they work with these questions with merely awkward words.

Knowing that adults struggle to gain clarity, can we even speak about these realities with children? Adults usually want a precise answer, a conceptual explanation from a certain world perspective or maybe just a theological, dogmatic formulation. But all of that is "stones for bread" when it comes to children. First of all, they do not understand it. Secondly, it is not what they are looking for. They want answers, but that does not always mean in words.

It means in something real, something they can know and experience with their whole Self.

If the adult has resigned and prioritized the daily duties, given up on having a closer relationship with larger realities in life, this adult cannot give the child at this age what he needs. Children refuse to resign. They have so much life in themselves that they strive, no matter what, to understand the larger relationships in the world.

Children need adults as authorities on whom they can count. Yet they want, especially as of the tenth year, to notice that adult authority serves something much larger, something that lives both out in the world and in their Self. Whatever adults say or do is all measured with this yardstick by the children. Does the adult have devotion for his teaching of nature, of flowers, animals and the stars? Or does he patter out the facts in order to imprint them on the children? Or maybe he merely airs his own opinions.

If the adult has worked in such a qualitative relationship with the starry heavens as set out earlier in this article, he may have a path from which to present astronomy lessons in a way that serves children. Every star, every planet, the finest details in all of nature awaken devotion, something that is real for children.

The adult's perspective on the world, no matter how good and correct it may be, is always indigestible for children, something foreign and irrelevant. The same is true of merely naming the names and numbers of stars, planets and their movements. That is also something foreign to children, something they have no use for in their human development. On the other hand, if we present the names and numbers in such a way that the children can experience the qualities of the stars, planets and movements, the children can wrestle and live with the material in a valuable way. The very simple, observable things are important to work with, to weave them into pictures that are understandable and thereby digestible. Nature becomes a picture and, at the same time, the soul-spiritual in the child, that could easily become lost in the unapproachable, becomes condensed in a perceptive picture like the butterfly that emerges from the caterpillar. When such teaching takes place, children are helped to take a step further in their development, to finding their Self even better than before, to uniting their Self with the world.

Teaching Biology in a Human Context

by Graham Kennish
(Reprinted from *Steiner Education*, Vol. 22, No. 1)

"Your body is a space capsule, your head the command module." So begins the introduction to a three-dimensional moving pop-up picture book on the human body now available in the UK. "When you reach puberty your hormones switch on," announces a heading in the London Science Museum permanent exhibition called *A Study of Ourselves*. An advertisement for beer displayed on billboards in the UK recently shows a series of ape-like figures progressively reaching a vertical posture, the penultimate figure with a bowler hat (symbol of the English business gentleman) and the final figure carrying a can of the appropriate beer. A question mark points to the potential evolutionary leap which awaits discerning drinkers.

These three examples are particularly gross reflections of deeply held beliefs in the West, beliefs firmly underpinned by faith in scientific objectivity. One of these is that the human body is nothing more than a highly complex machine which human beings will eventually be able to take apart and reconstruct. A second, that our bodies and our minds are subject to the outcome of a complex chemistry. The third, that human beings have evolved from a primitive animal condition and that any further evolution is in the random and arbitrary hands of environmental influences. In teaching any science to adolescents, one is aware of the forceful nature of these beliefs which are carried subliminally or openly throughout our culture.

One of the hallmarks of a good scientific theory is that it should be capable of being disproved. This would seem to guarantee the absence of dogma in science, as any theory worth its salt will, by definition, eventually be superseded. Human nature, however, is stronger than scientific principle, so from black holes in space to human evolution, theories rapidly harden into tablets of stone brought down from a mountain of research. If we experience surprise, displeasure or vague discomfort in reading such statements as those below, then we can be sure that we are taking current theories for granted or carrying memories of school science unchallenged within us:

- atoms do not exist,
- human beings did not evolve from apelike ancestors,
- life did not arise from a primeval organic soup, nor the universe in a gigantic explosion,
- the sun is not a ball of atomic fire, the heart is not a kind of pump,
- the brain is not a kind of computer.

Any twinges? How free are we to consider alternative views or even challenge current ones in our thinking? Cemented and consolidated over years of experimental measurement, poured out in textbooks, magazines, films, games and models, opposition can invite ridicule, disbelief, or the accusation of a lack of objectivity or ignorance of highly specialist research techniques. Yet the history of science is filled with the birth of ideas that ran directly counter to customary modes of thinking, and the very birth of scientific inquiry was the birth of an independent and free spirit of inquiry, unconstrained by tradition and religious or social pressures and prejudices. The biographies of Galileo and Darwin illustrate the struggles.

What is objectivity? Is it confined to what can be measured in mass, distance and time, or can it include the faculty of observation, thinking, and an open mind? There may be few technological or military applications in an open-minded contemplation of the universe, but this must surely always remain the bedrock of free-thinking inquiry and scientific progress. We stifle or undermine it at our peril.

These considerations are crucial to the teaching of science to adolescents in a Waldorf school, and the main lesson periods in Human Science for Grades 9 and 10 (14- to 16-year-olds) illustrate this. The periods are most commonly known as Human Science to allow the widest possible context to the biology arising from this. How can the wealth of knowledge currently available about the human body, for example, be presented without a fragmentary succession of organs and systems, implying that all these and more constitute the whole? What meaning does the liver have if separated from its context of blood and digestion?

What meaning does the digestive system have if not in the context of daily life? What meaning do any of these organs have in relation to my inner experience as a human being? A teacher could be tempted to overcome these problems and make the subject relevant to life merely by linking the study of items of popular interest: alcoholism, liver disease, digestion and a healthy diet. Such connections stimulate interest but alone they fail to meet the adolescent's deeply held conviction that there is meaning and mystery in the world. Unless there is a context for biology which carries "food" for these, an adolescent's thinking will be confined by the practical or the popular, and deeper, less conscious questions will remain unaddressed.

One aspect of biology that has, until recently, received little consideration is that of form—why living things have the shapes they do. In human science a considerable part of the work should be an experience of form through drawing and modeling, so that observation and thinking retain all the mobility which other lessons like those in music, art, and movement have developed.

To illustrate how observation and thinking can create a meaningful context within which fine details and invisible processes can be studied, take the form of the human skeleton. The form of the human head is spherical. At the extremities of the body are opposite forms, linear, angular, jointed. Between head and legs, the ribs (which are both linear and curved) create a materially incomplete enclosed surface (the rib-cage), protecting the soft tissues of heart and lung as the skull does brain and the bone, marrow. The head is still, the skull plates fused, with only the jawbone movable, like a limb. The upper ribs move in breathing but are fused together by the sternum. The lower ribs are freer, hanging limb-like, opening up the chest cavity into the abdomen. Following these forms and their movements, the opposite qualities of skull and limb with the balancing features of the ribs are clear to see.

Care in observation and a thinking faculty which can rove over the contrasts as well as remember details and named parts build a meaningful whole, not quantifiable yet not arbitrary or fanciful. Such a picture can be followed into the forms of individual bones (the 'head' of the femur, for example) before it fades into the detail of bony tissue and the process of ossification. This type of approach can be extended to other parts of the body as well as the systems of organs commonly considered in a quite different context in any standard textbook or encyclopedia. The fundamental difference is that relationships between organs and systems can be considered and grasped without recourse to theories about neural/electrical transmission, nerve/muscle reflexes or sensory/motor nerves. Such considerations should follow as elaborations of thought, not be considered the foundation for understanding the whole organism.

The nervous system is centered in the head with nerves leading to and from all parts of the body. The brain rests, partly floating free of gravitational pressure in the cerebrospinal fluid. There is no movement, impulses are silent, invisible—we feel most awake here, alert, our senses concentrated here. Below the diaphragm, a sheet of muscle which divides the trunk, the digestive system begins metabolic processes whose outcome supplies the energy needs of muscles to meet the demands of gravity in the limbs. Movement, warmth and activity prevail here, and as a complement to the senses which receive impressions from the environment, limbs reach out and impress themselves upon it. Between the two extremes of shape and activity of head and limbs (nervous system and metabolic system) are the rhythmic movements of heart and lung. Rhythm is movement and stillness in harmony. Breathing leads substance outward into the world and receives from it. The circulation of blood gives and receives inwardly.

Our experience of feeling is centered here, brought to consciousness by the head or expressed in movement through the limbs. Our inner experiences as human beings have their reflection in the form and activity of the bodily organs.

This brief and sketchy attempt to show the context within which the details of the human body may be taught to adolescents seeks to illustrate how a sense of wholeness and meaning can be the foundation of such a study. These pictures are neither fanciful nor arbitrary and are available to any keen observer with the controlled imagination which lies at the heart of objective knowledge about the world. They lead the adolescent to respect and have confidence in his own unaided faculties, so that further study of details and reading about experiments which explore the most minute aspects of physiology can be related to a meaningful whole.

Another message received is that knowledge about the human body does not rest solely on the biochemical or genetic analyses of experts, but is a mystery open to any keen observer with clear and mobile thinking. Adolescents also have a context within which to appreciate and admire the results of medical technology alongside the deeper issues raised that challenge human attitudes to birth and death. The adolescent's burgeoning inner life is also confirmed as a reality which the body supports and responds: I have a brain but I am not my brain. I have feelings but I am not my feelings. I have a body but I am not my body.

Another outcome which follows a consideration of form in the human body is a consideration of the balance and harmony in its architecture. Observation of the animal world can show very clearly that by a one-sided emphasis of the peripheral parts of the human skeleton, for example, specialized animal limbs arise. For example, an extreme development of the digits of the hand gives rise to a bat's wing, bone for bone. One-sided development of the forefinger alone creates the horse's hoof, while any distortion of the balanced forms of human teeth quickly creates herbivore (cow), carnivore (cat), and rodent (rat). Distortions of the human form always give rise to animal-like caricatures as the political cartoonist knows well. Such considerations leave open the question of how the human form evolved. Studies of the animal world return again and again to the human being without whom, after all, observations and questions would not exist.

In contrast to this emphasis on form comes what is usually considered the "real" content of biology: details of gaseous exchange in the lungs, respiration in the tissues, hemoglobin in the blood, excretion in the kidneys, enzyme activity in the alimentary tract. These substances and processes are not directly visible, and most of what is known about them is the outcome of detailed experimental investigations. They demand clear thinking and are vital exercise for the growing teenage intellect quite apart from the factual knowledge of their content. In differing degrees, with examples to stretch the ablest pupils, a whole class should experience the real satisfaction of understanding such biological processes and how each is coordinated with another, to create a harmonious balance against a changing environment. The context outlined previously makes it harder to fall into the satisfaction of the clever intellect which would anticipate that if only enough details were known and added together, like a gigantic biochemical construction kit, the whole organism would be explained.

As with all Waldorf teaching, the choice of what to study is so vast, that the question immediately facing a teacher is where to start. And so the choice is best determined through consideration of the developmental age of the class—also a Waldorf principle. In Class 9, thinking powers are usually not so fully developed as they will be in Class 10, and adolescents live very sharply in their senses. So a beginning can be made successfully with a study of the skin and the very visible and obvious sense organs, for example the eye. Its intricacy and sensitivity awaken and challenge any tendency to superficial comparisons with a camera. Analogies usually have very limited value and, when held to as teaching aids, seriously distort accurate observation, memory and thinking, leading quickly to the false sense that the eye has been explained. Details challenge such easy paths to understanding as these considerations may illustrate.

The light-sensitive cells in the retina actually point away from the light. The act of seeing involves the whole organism, not just the eye, and the image which reaches the retina bears little resemblance to our perceptions of the world around us. We cannot see light but only the outcome of its penetration of matter. At night, outer space is filled with light from the sun but appears black until reflected by the moon. So what is it that we "see"? Is it a coincidence that we say "I see" when we understand something? Dim stars cannot be seen looked at directly but appear when the focus of our gaze is turned slightly to one side. Is this not often true when we search our memories and thoughts? Suddenly a thought "dawns on us" and we see it "in a flash." The genius of language leads us to the widest considerations.

So, the opportunity arises to consider such fundamental questions as the nature of light or the biographies of individuals who have been deprived of sight or hearing. The widest possible considerations should be able to arise within a human science period.

In Class 10, the main lesson includes embryology and, as far as time allows, such themes as child development, racial and cultural differences, temperaments and personality. At this age a pupil's thinking ability and maturity are usually much more capable of doing justice to such topics, particularly to the development of the organs and the ethical questions that arise over abortion and embryo technology. The major organs of the body having already been considered, their development from layers of folding and enfolding tissue within a matrix of sustaining membranes is a rich and rewarding study. I have felt very privileged to experience how boys and girls of this age can be together and feel free to share some of their deepest concerns and feelings on life, death, and relationships arising from an objective study of this miraculous process of development.

At an age when sexual physiology and the "facts of life," as they are so inaptly called, are outwardly known and serious relationships have often already begun, this study of embryology can come as a healing force to adolescents in their struggles to cope with the freedoms and the responsibilities our society has placed on them. It can be a fitting end to such a study to also include the question of old age. While we talk easily of growing up, it can be quite a revelation to adolescents that to become an adult is to embark on a lifelong experience of learning and development. The "seven ages of man" need not be limited to a decline from youthful vigor to dull senility but may include the pathway to self-knowledge and wisdom. Herein lies an opportunity for adolescents to perceive that the excitement and the challenge, as well as the doubts and anxieties they are struggling with within, are shared by parents and grandparents, too.

Graham Kennish initiated a Waldorf teacher training program at the Wynstones School in Great Britain. A long-time Waldorf educator, his primary interest is in developing phenomenology in science teaching. He is currently a professor at the University of Plymouth.

Aesthetic Knowledge as a Source for the Main Lesson

by Peter Guttenhöfer
translation by Peter Glasby in consultation with Georg Maier

The main lesson of the Waldorf school is different from a double lesson. It is a unity of three parts, composed like a sonata.

The classical sonata form is three movements. The first movement, the "head movement," presents the theme, forms it out, turns it around or mirrors it, submits it to many dismemberments and distortions (*Verrückungen*) and, through the variations, makes the listener once more aware of the drama of these transformations. The second movement brings a totally new atmosphere, in a slower tempo and a changed key, however, still totally related with the musical "substance" of the first movement. Here the task is less a working through of the theme than the direct touching of the inner space of the soul. Finally, in the third movement, the restrained drive for movement is let go. Quickly, the rhythmically accentuated, thematically light-footed, final movement plays itself out. Here too it arises from what was laid down in the first movement, but it still has something of its own. Like something refreshing, the playful scherzo (joke) often slips in, so no main lesson may go by without the weight of the content, at least once, being lightened and relaxed by laughter.

The main lesson and the sonata are both artistic compositions in which the three sections go out organically from a middle point. They are not simply put together in an additive way, but arise out of transformation of the foregoing. Between the sonata movements are small pauses, but without interruption of the musical flow, instead a deep drawing in of the breath, a short repositioning of oneself, a thoughtful clearing of the throat. The applause and coughing come only after the final movement. There lies recess.

We begin to realize that these three parts, which build a whole, are somehow related to the three parts of the human organism, which is why the sonata is so healthy. The main lesson too should be health-giving, not only instructive. We can understand it as an aesthetic phenomenon, like the sonata. Rudolf Steiner spoke of the "art of education," and practice shows that the main lesson must be artistically formed. With this the question is raised about a kind of aesthetic concept for school teaching. The concept *aesthetic* is used for a process which is artistic and which has, as a product, a piece of art. How do I teach artistically?

The three sections of the main lesson are determined by the threefold constitution of the human being, in which the bodily division into the nerve-sense system (head), rhythmic system (chest) and metabolic-limb system are related to the differentiation of the soul in thinking, feeling and willing. This threefoldness is also related to the three steps of the logical process—conclusion, judgment, and concept—which Steiner (1919) presented in his ninth lecture of *The Study of Man*. Here, he reversed the usual Aristotelian logic and put the word *conclusion* in a provocative way, at the beginning of the logical operation. It does not mean conclusion in the sense of "end" and also not in the sense of *conclusio*. It does not mean that a thought process has come to an end and a deduction is being made. It refers more to the process wherein the human being and the world encounter each other, where phenomenon and sense perception meet or where the phenomenon appears through the world colliding with the human being without swallowing him or carrying her away in a sleep condition.

World will pushes up against the dark self will, which the human being carries out in his embodiment. The "I" touches, through the sense perception, those deeper levels of being, from which all phenomena press into appearance. And the human being does not fade away in that great fire; instead, because she experiences herself so closely connected to the world, she closes herself off, protects herself from becoming one with the world in the act of conclusion. Goethe expressed it:

> When I at last come to rest with the archetypal phenomenon, it is still only resignation; but there remains a great difference if I resign at the limits of humanity or within a hypothetical limitation of my narrow minded individuality. (Goethe 1817, Verse 138)

Yes, the conclusion is the amazing moment of phenomena emergence, before defined representation (mental picture), before the wandering judgment, before that hypothetical limitation. Steiner (1919) stated laconically: "The lion is a conclusion."

The judgment links itself to the conclusion, or, the act of the conclusion awakens the movement of judging. And at the end of the judgment—weighing up, comparison, affirmation, and so forth—stands the concept, which created the quiet in Goethe's soul. The fiery seconds of the conclusion stand in polarity to the constancy of the worked out, not misunderstandable word formulations of the thought form. Now that the long practiced, joyfully suffered syllogism has been overthrown, and so that the old meaning of conclusion does not shadow the "conclusion," the conclusion stands in the middle of the main lesson.

Something new from the content of the main lesson is presented, in the most various ways: A physics experiment is demonstrated, a historical event is described, a botanical drawing is observed, a new problem type from trigonometry is presented or literary text is read, and so forth. The teacher is active, the students take it in, silently. They do not write, they are totally sense organ. In this moment the pure inner will activity of the students prevails. The emerging appearance of phenomena is prioritized above all understanding. No question is allowed. The world touches the student, who lets him/herself be touched. The student becomes "world" and not only an "observer of the world." The student forgets himself and is totally "in" the thing (*inter est*, interest).

The total absorption in the Archetypal phenomena sets up in us a kind of anxiety: "We feel our inadequacy." Goethe (1817, Verse 137)

This does not mean that there is always something of the archetypal phenomena in the teacher's presentation. However, the emergence of a phenomenon has something fundamentally numinous, and the feeling of inadequacy awakens the need to judge, to take a position, to reject or to become enthused. So, after the teacher has completed his presentation, there begins the judgment and the third section of the main-lesson sonata is played. It ends open-ended and the students go with the opened-up and unsolved problem into the recess. The Waldorf teacher tries to take into account that in the coming night the perceived riddles are taken into the sleep. What that means is "withdrawn from our usual consciousness" and shall not be further explained here. What matters here is that, when the students appear in the main lesson the next morning, they are in a completely changed relationship to the content of yesterday. With quiet, almost serenity they now go with the teacher into the thoughtful business of working the concept to the phenomenon. That is the first movement of the new main lesson, which again is followed by the 'conclusion' event and the third movement of judgment. Each main lesson begins, therefore, with the concept part, which works with that which has arisen from yesterday. This gives the sequence: Concept—Conclusion.

Judgment

The logical cognition process on a topic, however, runs with the structure: Conclusion—Judgment—Concept. The night is taken in between Judgment and Concept. In this way there are always three days of main lessons following each other that belong together. When the teacher forms the main lesson with this in mind, there lies within it a spiritual dynamic of its own.

From where does the nourishment of this process proceed? It comes from the event in the middle part of the main lesson, from the encounter of the student with the reality of the world, not from a speaking about a somehow imagined reality. Everything depends on whether or not an actual connection (*schliesseri*)—a happening (conclusion)—happens for the student. Out of this insight arises the task of furthering the concept of "aesthetics" in relation to the main lesson. That it is related in its structure to the musical sonata makes it, when successful, an artwork. However, the actual aesthetic process is grounded in the being of the conclusion.

To understand this, it is necessary to release the concept of *aesthetics* out of its traditional frame of meaning. There have been a series of researchers who have tried to do this in the last few years in connection to A.G. Baumgarten's *Aesthetica* (Baumgarten 1750/58).

Wolfgang Welsch (1990) described how, in the time after Baumgarten, "there was a restriction of the concept of aesthetic predominantly to art or even to only what was beautiful. That, in my opinion needs to be turned around today." (Welsch 1990, p. 9). Especially the work of Hans Rudolf Schweizer has brought recognition of Baumgarten's original aesthetic concept and paved the way to an understanding of aesthetics, not as a theory of beautiful art, but as a philosophy of sense experience. He formulated Baumgarten's fundamental principles into the language of our time in the following way:

1. Aesthetics is not a specialized area within the whole of life's process, but the basis for the experience of reality.

2. Aesthetics brings the unbroken phenomenality of "things" to validity. It is, as pure phenomena, the unrepeatable, individual happening in time.

3. Aesthetic cognition is a purely intuitive cognition, which at first remains without conceptual treatment. It is that knowing on which one must rely in daily life.

4. Aesthetics is a field of relationships between the human being and the world, subject and object. If one denies it having any objective meaning whatsoever and ascribes to it a simple subjective feeling or a subjective "forming power," then one has lost its content and its being." (Schweizer 1976, p. 74)

With this we understand that the happening of the conclusion is the moment of aesthetic experience (Barth 1999, p. 111). Here there is no limitation to a specialist area; this is not only about the observation of art! Here we glimpse the existential moment of world encounter in the pure perception. We are standing at the spring for all teaching.

In reality, can there be such a moment in the practice of a school? Is not each pedagogical activity narrowed to the discursive symbolism of a science-orientated theoretical cognition, which limits necessarily "the life" out of school—something, which in fact all the students of the world experience when they are older than twelve years? The concept of *conclusion* as the moment of the aesthetic condition (Schiller 1793/94, Letters 20 and 21), in which the world and human being stand before each other naked, really means that just in the center of the lesson, "life" touches the student in the most intimate way, much more strongly and purely than in ordinary existence. This ordinary existence presents itself mainly simplified for trivial aims or modified through desires of all kinds and is seldom presented unclouded. Generally, the everyday person goes on his way in a fog in regard to the meaning of things. Actually, more than that, he does not even know the names of the plants which grow in front of his door. So then, this center of the main lesson is always a special space in which the beings of things can show themselves: the shiny silver pearl of molten tin, a quince leaf, the description of the sea battle of Salamis, or the sudden illumination of the connection between the pentagram and the Golden Mean. The objects of teaching are not won from conventional ideas of a canon for general education, but from a sense for the "symbolic meaning" (Cassirer 1982, p. 235) of things or processes, which as archetypal phenomena of the experience of meaning can speak—in a more fundamental way to the sense cognition of the student.

If such a demand is hard enough to fulfill in a natural science lesson, then lurking in the humanities subjects are even more awful conditions, which threaten to lure the teacher off track. Chief among these conditions is the opinion and the longing that everything must be "interpreted." What is the meaning of Hamlet? To this question, there can be as little a satisfying answer as to the question of the meaning of a mountain stream after a thunderstorm (cf. Schadewald 1974, p. 206). Yes, but how can we then read one of the greatest tragedies of antiquity, with our students and allow them to experience directly what Holderlin tried to express in the words:

> The presentation of the tragic depends mainly on the unspeakable, of how God and the human being are paired, and the boundless nature power unites in rage with the most inwardly human, thereby understanding that the boundlessness becoming one, purifies itself through boundless separation.

The heart piece of our 11th grade main lessons, Wolfram von Eschenbach's *Parzival*, reveals most clearly what is actually demanded. *Parzival* is not a book about the Grail. In fact, it is not a "book" at all, as the author emphasizes: "Who wishes to hear of further adventure, should please not take it as a 'book.'" (Wolfram von Eschenbach, verse 115, 25-116, 4). The reading of this book is itself an approach to the Grail.

Rudolf Steiner said: "No one gets near to the Grail with words or indeed with philosophical speculation. The Grail is approached, if one allow, all these words to be transformed into sensibilities (*Empfindungen*)." (Steiner 1914, p. 109)

The transformation of words into sensibilities in the soul of the teacher during his/her preparation allows a process to begin, which allows reality to emerge for the student. The student communes with this reality in the "conclusion" happening of the main lesson. Then finally, comes the scene, in which Parzival redeems the suffering of Amfortas, with the question: "Uncle, what ails thee?" (Wolfram von Eschenbach, verse 795, 29). This question is the archetypal phenomenon. To it there is no answer. It heals directly.

References

Barth, H. *Erscheinenlassen*, Basel: 1999.
Baumgarten, A.G. *Aesthetica*, 2 Bde. Frankfurt/Main: 1750–1758.
. *Theoretische Ästhetik*. Die grundlegenden Ausschnitte aus der "Aesthetica," Hamburg: translated and published by H.R. Schweizer, 1983.
Cassirer, E. *Philosophie der symbolischen Formen*, 3.Teil, Darmstadt: Phänomenologie der Erkenntnis, 1982.
Eschenbach, Wolfram von. *Parzival*, Darmstadt, 1918.
Goethe, J.W.v. *Spruche in Prosa*, Stuttgart, 1967.
Holderlin, F. "Anmerkungen zum Oedipus," in: *Sämtliche Werke*, Frankfurt/Main, 1961.
Schadewald, W. *Einleitung zur "Antigone" in Sophokles'* Antigone. Frankfurt, 1974.
Schiller, F. *Briefe über die ästhetische Erziehung des Menschen (Letters about the Aesthetic Education of Mankind)*, Sämtliche Werke, Stuttgart u. Tubingen: Bd. XVII, p, 72f, 1826.
Schweizer, H.R. *Vom ürsprunglichen Sinn der Ästhetik*, Oberwil-Zug, 1976.
Steiner, R. *Christus und die geistige Welt*, 6. Vortrag, 02.01.1914, Dornach, GA 149.
. *Allgemeine Menschenkunde als Grundlage der Pädagogik* (Translated as *The Study of Man* or more recently as *Foundations of Human Experience*), New York: SteinerBooks, 2004.
Welsch, W. *Ästhetisches Denken*, Stuttgart, 1990.

Dr. Peter Guttenhöfer is a teacher with over thirty years of experience, a founding member of the Teacher Training Seminar at Kassel, Germany, and a musician. He teaches history and German literature, among other subjects, at the Kassel Waldorf School.

Adolescents
Their Relationship to the Night and the Senses
in Connection with Their Own Development

by Peter Glasby
Mt. Barker, Australia

This is a description of the workshop, "Today's Child, Tomorrow's World," which took place over three days at the Kolisko Conference in Sydney, Australia, July 3–8, 2004. The methodology of the workshop was to develop several strands simultaneously, by providing experiences which were then allowed to ripen overnight and developed further on the subsequent day.

In his many courses to teachers, Rudolf Steiner gave material which, if worked with, provides a rich resource for understanding the theme, posing new questions and providing practical insight into working with adolescents in the classroom. Some of the answers are astounding. For example, education is not all about providing "light" for the students but also "darkness." The beginning of the learning process is a "conclusion," not the end of the learning process. (Guttenhöfer 2004). There is a need to differentiate between *living concepts* and *dead concepts*, a critical differentiation for education. There are implications for the way a teacher arranges his lessons and for the way in which the day is organized for the students.

In this workshop we explored these topics using text material from Rudolf Steiner's work, experiments, biographies, group discussion and active participation. We attempted to work with the same elements we work with at school: the sleep life, the structure of the lesson, and the daily and three-day rhythms.

Day One

The following table is a summary of the rhythms which happen within our daily lives. To which rhythm of life does sleep belong?

Year rhythm	1 year	Physical body	Zodiac
Month	1 month	Life body	Moon
Week	1 week	Astral body	Planets
Day	1 day	Ego	Sun

The astral body is the name for that spiritual member of the human being, which, in a certain way, contains our thoughts, feelings and impulses.

The day rhythm of the sun is one of renewal, of starting again. The change from our waking consciousness to sleep consciousness and then back again to waking consciousness is one of the fundamental rhythms of our existence. Without one side of the rhythm, we would become ill and wasted; without the other we would lack the basis of experience for burgeoning independence. On behalf of the adolescent, who is undergoing significant changes in his or her own relation to his/her soul life, there are particular issues the teacher should pay attention to:

How is this change happening for boys and girls?
How is the sense life to be cultivated?
How is the process of knowing managed in school so that the students can be active participants?

In the first lecture of the series given to the first teachers at the first Waldorf school, we find the interesting statement: "We must teach the children how to sleep." Surely school should be about keeping children awake, even amused, but what on earth are we to understand about teaching them to sleep? This statement leads us into the question of what is sleep. What is this state in which we lose our memory, where we recover, reorder, and are healed?

During our waking consciousness, our senses are filled with the images of the world. But they are closed during sleep and our consciousness is partially extinguished. Steiner described what occurs then in our soul. Freed from the distraction of the senses, the soul is given over to a world of order and archetypes, out of which the creative forces of the universe flow.

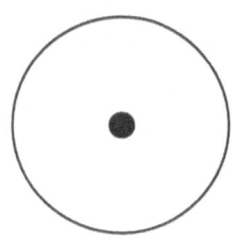

The symbol for the sun is a circle with a point in the middle. This is a picture of the human experience of the day rhythm. We are a point when we wake up into our senses, our day waking consciousness. At night we lose our consciousness into the world of sleep, the world of the dark, the stars and the periphery. What are the phenomena we can recognize in relation to our sleep life? It replenishes us so that we are ready to look at things afresh, to take up new tasks. We can go to bed plagued by a problem only to wake to find the solution has dawned upon us (literally). There is also a ripening that happens with sleep; skills and abilities mature to new levels. The closeness of events from the day appears in a new perspective, which is why we often sleep on a situation before we make a decision. We can consciously take a problem or situation involving another human being into our sleep and experience how, upon waking, much of the tightness of the problem knot has relaxed and new possibilities have emerged.

Teaching the children to sleep is also mentioned in Lecture 1 of *The Study of Man*. One could think of this as teaching the children to breathe between the periphery and the center. The medieval mystic Angelus Silesius refers to this aspect of the human existence between center and periphery:

> I don't know what I am, I am not what I know,
> a thing and not a thing, a drop and a circling.

This concept was depicted in medieval times in many different ways. One of the most beautiful is represented in Fig. 1 of Steiner's *Occult Science* (1925) in the chapter about sleep and death. Waking and sleeping are described as the astral/ego being within and without the physical/life bodies. The life body is form-giving but only if it receives from the astral body the pattern forms or archetypes. In wakefulness, we turn our senses to the surrounding world to form mental images. These are the "disturbers of the peace."

The physical/life bodies contain the organs by which the astral body perceives the external world and has to be separated from its own world. (See *Occult Science*, p. 316) In sleep the astral body is united with the world out of which the human being is born. Apparently people in medieval times experienced this world out of which we are formed as being the zodiacal starry world.

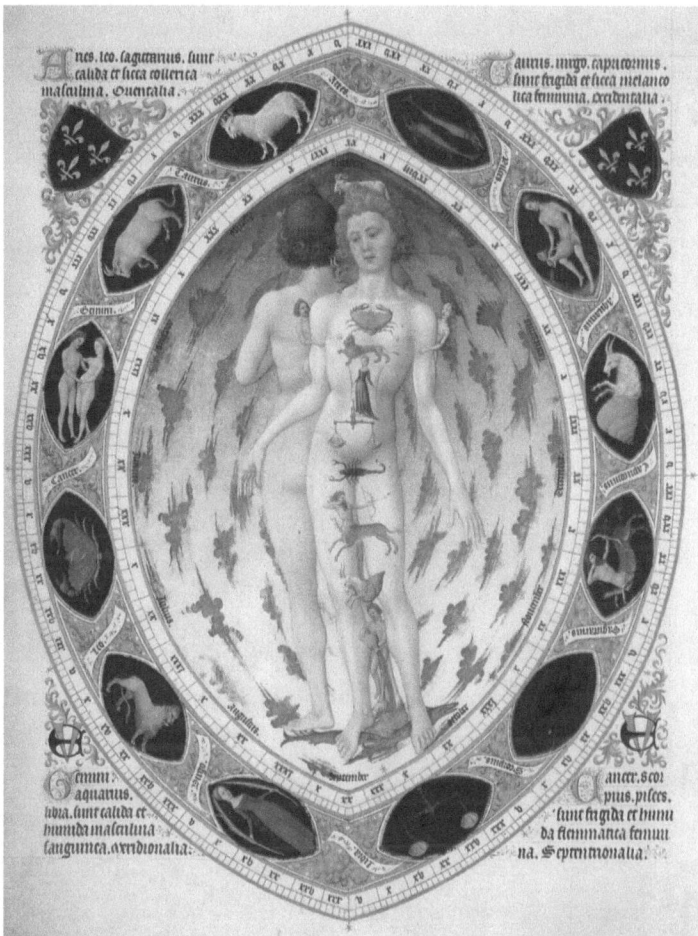

Fig. 1. The Zodiacal human being is a recurring theme in medieval manuscripts. The signs are associated with corresponding principles in the body to show the human being, the most perfect creature in the world, as the microcosmic image of the heavens—a reflection of the celestial mirror. (from MS. Les Tres Riches)

Through the senses our experience is dismembered among twelve fields of experience. In cognition this dismemberment can be remembered, reintegrated. How this happens is important for the art of teaching. It swings between the dismemberment of waking consciousness in the twelve fields of experience (twelve senses) and the reunification in the nightly starry world of twelve zodiacal directions of space. We can have an experience of dismemberment, but we cannot so easily experience the reunification within the starry world. We can, however, experience the extraordinary rejuvenating power of sleep.

Up to puberty the astral body is in gestation. Like the physical body in gestation, it is there, being nourished, but not yet born until adolescence dawns. The birth of the astral body is related to the development of judgment at many levels through the high school years. This synchronization is important for educators to realize. Goethe (1960) wrote about the space between the experience of a phenomenon and the judgment about it:

> One cannot give enough importance to the fact not to draw conclusions, not to prove things too quickly, or support a theory. Between the observation of the facts and the judgment, there belongs a space in time to take up the facts in a pure way. They must live. The mind has the tendency to jump to conclusions.
>
> – Goethe, "The Experiment as the Mediator between Object and Subject"

This has important implications for the way experiences are prepared for the students, the timing of the explanation, and the way of the explanation. This principle was a central theme in the practice and art of teaching by Steiner, in lectures of 1919 and 1921. In Lecture 9 of *The Study of Man*, he characterized the sequence: Conclusion, Judgment and Concept, whereby the Conclusion is characterized as a "happened event" and stands at the beginning of the process of cognition not as a conclusion concept.

At this point in the workshop, we did two experiments (from von Mackensen 1992). The first involved going out into the parking area and looking into a gray trough, full of water, with two tiles at each end, one vertical, one horizontal, with alternating black and white stripes painted on them. We tried to pay particular attention to the transition areas of dark to light.

The second experiment was conducted inside. We looked at a person through a large water prism, again paying attention to the transition areas between the light and the dark.

In the first experiment we noticed, after some time, that there was a shift of the image in the water. The closer we brought our point of view to the water, the more the tile seemed to be compressed toward the surface of the water. At the same time colors appeared at the edge of the stripe—warm colors on some edges and cool colors on other edges.

What we saw through the prism has been drawn in the picture above. Notice the colors on the edge of the dark and light areas. These experiences were described and somewhat characterized, and then left. To conclude our first day the following story was told. (Davidson 1965)

In the 1860s an Irish couple, Alice and Thomas Sullivan, escaped the famine in Ireland and settled in a little farming community in the eastern United States. Their first child, a little girl, Annie, was born on April 14, 1866. Her first two years were blessed and then her life became filled with hardship. When she was two years old, her eyes became itchy and her parents, using the poor man's doctor, Time, waited for the condition to go away, but it did not and was finally diagnosed as trachoma. Soon after, her mother, Alice developed tuberculosis and the next child, Jimmy, born in 1869, had tuberculosis of the hipbone. Thomas began to drink. A third child was born. Annie became a difficult child with frequent, violent temper tantrums. One day her father said to her after she had thrown his shaving gear all over the bathroom: "Are you a devil? See what you've done. Brought bad luck to the house. Seven years of bad luck."

When Alice died, the children were all separated amongst the relatives. No one wanted Annie because of her tantrums and because by now she could hardly see. After an unsuccessful stay with a cousin, Annie and Jimmy were taken to the Tewksbury poorhouse. Jimmy was only allowed to stay with his sister by wearing an apron. They lived in appalling conditions; their playroom was the room where those who died were kept until burial.

Then Jimmy died, and Annie's only friends were two old women, one of whom was blind and the other arthritic. The blind woman told her stories and the arthritic woman would read to Annie from magazines in exchange for little chores. A year later, a visiting priest, Father Barbara, noticed Annie and took her away to the Sisters of Charity Hospital in Lowell, Massachusetts, for an eye operation. This was unsuccessful and Annie was returned to Tewksbury where she continued to live in deplorable conditions for three more years, 1878–1880. During a government inspection of the poorhouse in 1880, Annie ran amongst the inspectors as they were about to depart and cried out that she wanted to go to school.

A few days later, a coach arrived to take Annie to Perkins Institution, a school for the blind and deaf. She was now fourteen years old and had had virtually no schooling. She was difficult with the other children who called her Big Annie. She learned both signing and Braille quickly but could see no point to spelling. Her stubbornness and the unkindness of one of the teachers led to another serious tantrum and her leaving the room—and nearly the school. One of the teachers, Miss Moore, took on the responsibility for Annie, setting aside time every week, during which the two studied or talked or walked the grounds. Over time Annie began to heal. She began to imitate Miss Moore's soft voice, her gentle ways, and her kindly interest in other people. Slowly the manners she imitated became part of her, and the other girls began to warm to the new Annie. It was the beginning of a new experience, one she had not had for so long that she had forgotten it, the feeling of happiness.

Two other people contributed to Annie's further development. One was a young Irish man whose rooms Annie cleaned. He convinced her to see a Dr. Bradford who eventually convinced Annie to attempt a series of operations, which finally gave her back relatively good vision. Annie was one of the earliest patients to have ether as an anesthetic. The second was Mrs. Hopkins, a widow from Cape Cod, who had lost a daughter of Annie's age and who became Annie's housemother. In fact she became a mother figure for her until she completed her education in 1886. There were eight graduates in that class and Annie led them all. She was the class valedictorian and at graduation gave a speech memorable for its universality:

Now we are going into the busy world to take our share of life's burdens and do our little to make the world better, wiser and happier... Self-development is a benefit, not only to the individual but also to humanity. Every person who improves herself is aiding the progress of society, and everyone who stands still is holding it back.

Annie was dreading what life would now bring her after graduation. As she contemplated an uncertain future, the principal of the school brought her a letter which asked if she would consider taking a position as governess for a little blind, deaf, mute girl from southern United States. The girl's name was Helen Keller.

Day Two

The question arose about scientific language and the appropriateness of its use. This led to the re-consideration of "dead concepts" and "living concepts." Scientific language can carry implicitly dead concepts within it. Sometimes these unconsciously-carried concepts are brought to consciousness with application and then they may shock us. This type of experience is becoming more common as the applications of biotechnology become more prevalent in society.

On the one hand it is necessary that children learn the names of things and have knowledge about the actual nature of the world: how high mountains are, how deep the sea is, how long rivers are, what time it takes to travel between places. In learning about rocks and volcanoes in class six, one may well learn some of the language of geology such as igneous rocks, plutonic rocks, sedimentary rocks, and so forth. No real model is implied; the names are descriptive.

An example of inappropriate language is the language of rays and particles in explanations of brightness and shadow in classes 6, 7 and 8. In the light ray example a model is implied. Instead, it is better to develop your own language as in the cases of Von Mackensen (1992) and Maier (1986).

We read a section from Chapter 8 of *The Study of Man* (Steiner 1919) about the senses and the polarity of disparateness in the day waking state compared to the unity of the night state. This gave us indications about how to work with the night and the process of judgment. The integration of sensory experience is an important part of the process of judgment, and Steiner described how it begins in the way that sensory experience is integrated, as in the example of looking at a shape with color. There the sense of sight and the sense of movement are integrated. The senses give us a separate experience of existence; in fact many separate experiences, which, Steiner suggested, have twelve sensory fields. This disparity of the twelve senses and the unity of the archetypal human being represented by the twelve zodiacal signs forms an interesting polarity.

Night "I"

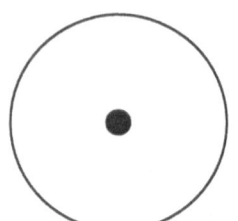

Unity of the twelve zodiacal signs as representing the archetypal Human being

Day "I"

Disparity of the twelve senses

We tried integrating the two experiences from yesterday, the experiment in which we looked at the striped tiles in water and the experiment in which we looked through the prism. In summarizing the experiences we made the following list:

- Colors appeared on the edges of the light and dark.
- Different colors appeared. On one edge there were warm colors, on the other there were cool colors.
- There was a shift in the image under water. The closer the eye was to the trough water surface, the greater the shift and the more intense the color.

The connection of these phenomena was not yet clear so we performed a new experiment to help relate the different experiences above (von Mackensen 1992). A glass tank was filled with water and placed high up for all to see, in front of a white board. Dettol or Savlon was added to the water and clouds of turbidity formed. These took on a warm, reddish color. Next a black board was placed behind the tank and another over the top of the tank. Now the clouds of turbidity took on a cooler, bluish hue.

In the second experiment, I filled the second prism with water and started by putting my face up close and then backing away from the prism. My view to the

audience was like this. When I was close to the prisms, my image was pulled out in two directions. As I moved backward from the prisms, the two images of me became even more separate. The letters *a b c d e* were drawn on a black board and placed at different distances from the prisms. The results from these experiments were recorded as below. There remained a lot for us to integrate on day three.

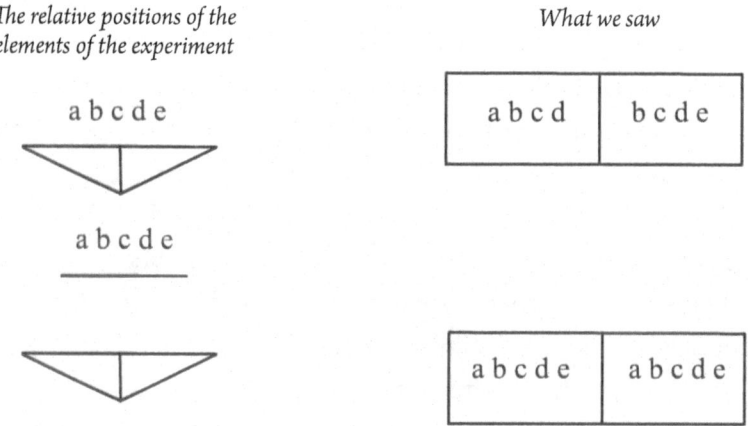

Day Three

We began with the story of Annie Sullivan coming to work with Helen Keller, the little girl struck deaf and blind by scarlet fever when she was eighteen months old. Plunged into a world of darkness and silence at the developmental time when as children we learn to stand and walk, then speak, then think and become self conscious, Helen was surrounded by sympathy as a prevailing soul mood. Her parents, unable to rise above their sympathy, spoiled the little girl and allowed her to do things that were out of control. Her behavior became so wild and uncontrolled that there appeared no alternative than that Helen be sent to the state asylum. There was one last chance and a letter was sent to the Perkins Institute for the blind and deaf. This letter reached Annie Sullivan and so the two were brought together for an ego encounter that I believe can be helpful in our understanding of the senses. There is the famous fight over the sausage when Annie refused to allow Helen to pick food off her plate. Then came the ultimatum from Annie that she could only help Helen if she could have her alone in a house for a period of time without the intervention of her parents. God only knows what happened in that house during that time, but we do know that, when the two emerged after some weeks, Helen had experienced an awakening that came from her meeting with her teacher, Annie Sullivan, an encounter that had not only been one of sympathy but also one of antipathy.

This primary encounter of another ego was seminal in the subsequent awakenings of the other higher senses of thought and word, which are remembered so well in the story of Helen being taught to sign the alphabet and even spell words. Although Helen was able to sign many words proficiently, she had no idea that they had any meaning. Then on a fine spring day, April 5, 1887, teacher and pupil went for a walk and stopped at the pump house. While Helen held her hand under the flowing water, Annie signed into it: W A T E R. Suddenly Helen's face lit up and shone with the dawn of understanding, then she dropped to her knees and pounded on the ground demanding to know the name of this too: E A R T H. For the first time in her life Helen realized that the signs on her hand had meaning. Within minutes she learned another half dozen words. Then she thumped her own head; she wanted to know her own name and then that of her teacher. At last she had realized that she was someone, and with that dawning came the experience of the other.

The stories of Annie Sullivan and Helen Keller are a wonderful instruction of how, amidst the tragedy of an unfolding destiny, a warm human encounter with even one person can make a life-changing difference. Secondly, the perception of the other ego is one in which we participate. This participation is twofold: sympathy and antipathy. Helen's parents were not able to perceive the ego of Helen; instead she was regarded as an object of sympathy. Annie on the other hand, was able to perceive Helen's ego and this expressed itself in the antipathy that was part of their encounter.

In this perception of the other ego, something was allowed to live in the world that until then had not been able to. In light of this story, it is interesting that in Lecture 8 in *The Study of Man*, Steiner gave his description of the twelve senses, starting with the ego sense in great detail and only then going on briefly to the others. What can we learn from this? In perceiving, we are helping to develop the perceived. In that we gain knowledge of something, we perhaps change the development of the beings interacting.

> When we look up to the wonder of the starry world, when we contemplate the whole process of the universe with its glories and marvels, then we are led at last to the feeling that all the glory that lies open to our view in the whole universe that surrounds us, has meaning only when it is reflected in the admiring human soul.
>
> – Goethe, quoted by Steiner in Lecture 1 of the series called *The World of the Senses and the World of the Spirit* (Steiner 1947)

Could it be then that the sense activity is something much more active and creative than the passive, physical process that is often imagined? The integration of the sense activity, which Steiner was already demanding early in the 20th century, is now part of the training for occupational therapists (personal communication with a former student studying occupational therapy). Below are some of the connections and qualities that were mentioned during the workshop in regard to this question. The twelve senses that Steiner (1966) suggested are listed in three groups with lines connecting them to another sense. The lines connecting the Body senses to the Spirit senses are interesting from an educational point of view because they point to transformational potential (necessity?). The content of the senses can have something to do with oneself, or it can have something to do with telling us about something else.

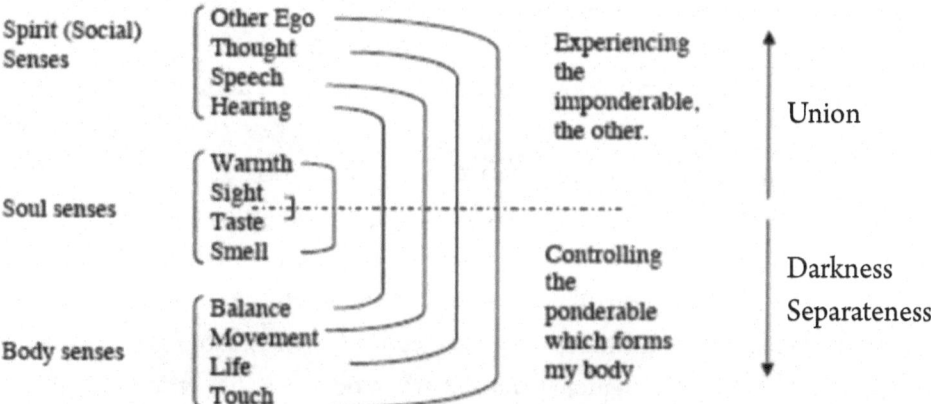

As teachers we need to be aware of these connections and integrate the sensory experiences into our lessons.

Particularly in a subject such as physics, it is easy for the lesson to race away in the sense of thought in regard to the mathematical laws of physics. However, as many students get lost in abstraction, it is beneficial, even necessary, to be aware of how the laws of physics are found in the human body, and it is there that they should be experienced first before they are made conscious as thoughts. A friend and researcher at the cutting edge of solar energy research calls this "body learning." Using Steiner's expression we could say that learning should be based in the body senses and take place over the whole human being.

In the workshop we began with the experience of seeing the warm and cool colors along the edges of the light and dark surfaces as we looked into a tub of water. This experience was simultaneous with the experience of seeing the image shift in the tub. The image is lifted. But in talking about the experience on the subsequent day, it was clear that we kept our two experiences separate. There had been no integration of the experiences. It is scary to think about how much of this can go on—where there are many experiences in a lesson which may remain unintegrated for the student. In Waldorf schools our goal is for the students to integrate their experiences. The teachers do not spoon-feed the integration, but we give lots of opportunity and support to the students to do the work.

On the second day of the workshop we managed to be clear about what was seen but not about the connections between what was seen. As a workshop facilitator I realized two things: (1) It would have worked better to give the experience of the conditions for the colors on Day 1; this would have made it fairly easy on Day 2 to come to the connection between the experiences; and (2) Given the situation I had created, the best way forward for Day 2 would have been to come to clarity about what had been seen and about what was still unclear. A whole lot had been learned: what way the image shifted in relation to the geometry of the prism and the surface of the water, that there was something interesting going on between the image shift and the color formation, even how to aim an arrow if you wanted to shoot a fish for dinner. But we had not connected the phenomenon of image shift with that of the colors. So now that we had reached the third day of the workshop, it was time to make the connections (the act of judgment) and also try to develop a living concept of what had been experienced.

 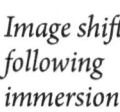  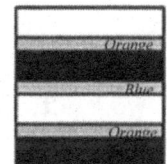

Image shift following immersion

Touchable visible tile   *Visible non-touchable tile with color bands*

We had discovered from the experiment on Day 2 with the Dettol in water that the turbidity in the water created conditions that were different in relation to light and dark. When the turbidity was in front of the white board or a light window, it darkened the lightness and warm colors appeared. When, however, the turbidity was in front of a black board, it had the opposite effect: It lightened the darkness, creating cool bluish colors. This is apparently what is happening in the separation of the touchable and visible image, both in the tile-in-water experiment and in the water-prism experiment.

As the image shifts up, where the dark bands move across the light bands, warm colors are formed, and where light bands move across dark bands, bluish colors are formed. This was what Goethe called the archetypal phenomenon: Bluish colors appear when darkness is lightened and reddish colors appear when light is darkened. We are surrounded by this lawfulness everywhere, in the blue of the sky and the warm glow of the transitions from day to night, in the way that a smoke from an outside fire appears bluish in front of a dark forest but reddish in front of the clear sky.

Here we have the test of a living concept. It does not end with its definition in the lab, allowing us a smug feeling of self-satisfied knowledge possession, but, instead, renews our interest to find the underlying laws again and again in the many conditions that appear in the living world. We had also to take the additional step of seeing where the image shift that we had seen in the prism led us. Here too there is a step of integrating different sense experiences. A series of drawings helps in the process of understanding the phenomenon. On the one side there are drawings from the plan view, which show how the elements of the experiment are moved in relation to each other; on the other we have drawings showing what was seen. Again we have an integration of two types of sensory experience, one from the tangible, moving world and the other from the visible world.

From these two results, we can see that the further the object abode is placed behind the prisms, the further the shift of the image. So we have a multiplication of images! How can we develop a system that not only multiplies images but also magnifies them? In other words, how do we go from prism shift to magnification? To do this we realize that in the image shift, each point was shifted by the same amount toward one pulling edge of a prism. However, if we looked through the window of the prism with the wider angle pulling edge, then the shift is increased.

In our example above, the following diagram represents the image shift. Each shift is by the same amount.

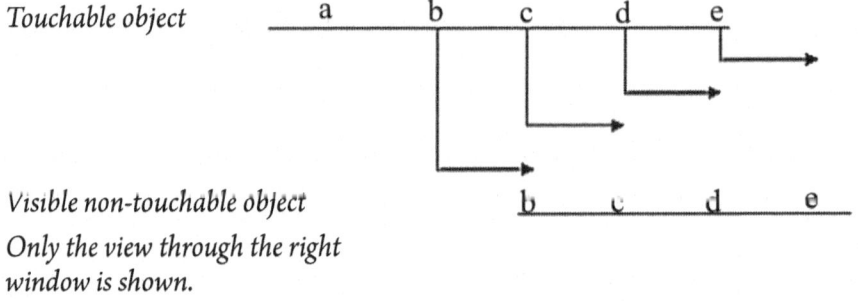

Touchable object

Visible non-touchable object
Only the view through the right window is shown.

Now what do we need to do so that each element moves proportionally to its distance from the middle point c?

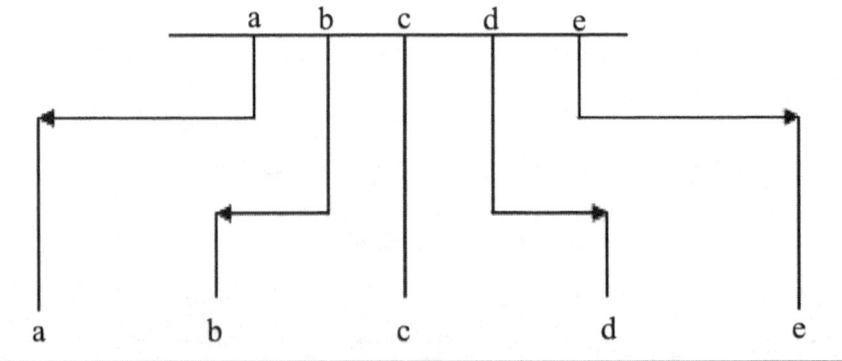

To get this magnified image we would need (1) the central point not to shift as in *c*, (2) *b* and *d* to shift a little, and (3) *a* and *e* to move the most. This would require a series of prisms with no pulling power in the middle (window glass) and an increased pulling power toward the edge.

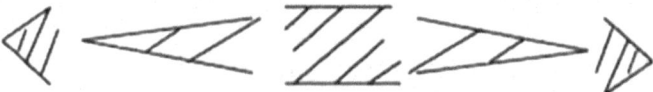

Putting this together would give an arrangement like this:

If this were ground smooth, then we would have a cross section which would be the cross section of a convex lens:

Finally we have come to an understanding of the lens, which can lead into a whole world of new phenomena. In the context of our workshop and being the last day, we experienced a series of new phenomena around the theme of the eye. Hopefully our detailed working beforehand had awakened our ability for integrating these phenomena without a great deal of explanation.

Experiment (von Mackensen 1992): The "chaos point" of a lens. The photographs below show the changes as seen by the workshop participants as the lens is moved away from my eye. The image of the eye is magnified as the lens is moved away from the eye until it reaches the "chaos point," when the lens seemed to be filled with the central point of the eye. This is also the point (i.e., the distance from the lens) when the lens is filled with the sun, as seen from the focal point, which then begins to burn. When the lens is moved beyond this distance, then the image becomes inverted and small—as though the lens goes beyond infinity in the point and returns from the periphery inverted, in a way that is similar to a conic section in the practice of projective geometry.

Finally a model of the eye was demonstrated using a flask in a special way and a convex lens study to one side (Maier 1986). The "chaos point" or focal point of the lens is at a point just short of the opposite face of the flask. A hazy image of what is in front of the lens is visible. (It gives an impression of what Annie Sullivan's limited vision was like during the time she was living in the poor house.) However, when the flask is filled with water, a clear image appears on the surface of the flask opposite the lens.

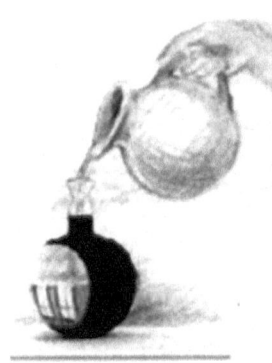

The lens of the eye is the whole eye. The eye too works between center and periphery. It is perhaps the key sense in the sleep process. After all, we close our eyes to sleep. There in the eye we also have the place where we meet ourselves in the other. This has been characterized so well by Ysaye Barnwell in her children's book, *No Mirrors in My Nana's House,* when she describes how a little girl finds her self-worth in the eyes of her Nana.

There are no mirrors in my Nana's house…
No mirrors to reflect the cracks in the wall,
The clothes that don't fit, the trash in the hallway.
No mirrors.
But there is love.
The beauty in this child's world is in her Nana's eyes.
It's like the rising of the Sun….

There in the dark pool of the eye called the pupil, separated from the white of the eye by the color of the iris, we find there, where the light of the other gleams forth, a source for our own self-knowledge. There in the dark pool of the night, separated from the light of day, we find the connection to our peripheral self. This is the secret of the indications given by Steiner to the pioneer teachers at the first Waldorf School: "You must teach the children how to sleep."

How can we now put this into our day-to-day classroom work with the children? Steiner described a method in his 1921 course for adolescence, Lecture 3 (Steiner 1965). In the course of the workshop we have tried to practice this method and integrate it with the other indication to working with the night given in his 1919 course to the teachers, Lecture 9 (Steiner 1966). There he characterized the steps of Conclusion (*Schluss*), Judgment (*Urteil*) and Concept (*Begriff*) in this unusual order, specifically beginning with the Conclusion. This conclusion is like a biographic moment, a happening that we witness and is then completed. That is the beginning of the way in the classroom: to go from an event that engages the whole human being, the whole of the lower senses, the middle senses, but without isolating the judgment and conceptualization from the process. Guttenhöfer (2004) has also written about this eloquently.

The Judgment is what is engaged between the biographic impact of the Conclusion and the Concept that is formed. The Judgment is engaged in the feeling life, in comparisons. It must be trained not to fall too quickly, like a bolt from the sky, but in silence and in the night to ripen and then be met in the classroom through questions posed by the teacher out of their contact with the students. The well-prepared questions strike a chord with the students and engage them. It is the teacher's task to find the latent questions that live often unarticulated in the feeling life of the adolescent students. They lead the biographic event, the Conclusion, after a night's sleep over into consideration and the forming of the Concept.

This is the third movement of Guttenhöfer's sonata. What has gone before resolves into a renewed interest in the world, a new connection to the world. By picking up the themes of the biographic event and connecting them to wider and wider circles of phenomena, the Concept that is formed is given shape and meaning but not in the sense of finishing the process of cognition but by leading to new cognition. This is the sense of the "living concept" as opposed to the "dead concept." In summary:

Day One

Experience—as sense-rich as possible and allowed to happen without commentary or explanation. This is not a time to spread light on everything. This is the Conclusion. It may be an experiment or the telling of a historical event. The whole body should be engaged. This does not mean doing gymnastics.

Characterization—What has been experienced is recalled as though by a witness for the whole of society. In the case of an experiment, the apparatus is cleaned up and put away and the event is recalled. The enthusiasm comes not from explanation but from clarity. The feeling life is touched and the children can be let out for their recess break. The darkness of the unknown is intimated. Expectancy grows. This forms a healthy beginning to the process of judgment.

Night/Sleep—The events of the day are integrated into the wholeness of the individual biography. Descriptions of the experiences of the lesson may be written up in homework if they have not been done during class. Connections are made at an unconscious level; latent questions arise.

Day Two

Questions—prepared by the teacher but also coming from the students. A mood of antipathy prevails in the room. This is not to be confused with unkindness but with demand. Here the teacher is in a position to require something from the students, that they remember what occurred the day before and try to answer the riddles that come from that experience. This can lead to debate, to feeling the vulnerability of uncertainty. Disparate experiences are brought to integration and then connected to the real world. A living concept can be formed: A new Experience—a new Conclusion—Characterization.

Night/Sleep—Descriptions of the experiences of the lesson may be written up as homework if they have not been done in class. Questions set by the teacher based on the morning's discussion may also be explored.

Day Three

Questions ... and so on. As teachers, this is the medicine we have for the young people of today. This is the daily work of the main lesson that can go from day to day, year to year. It forms one part of the day in the Waldorf school, the part that is most active in fulfilling the prerequisite for freedom.

BIBLIOGRAPHY

Barnwell, Y.M. *No Mirrors in My Nana's House,* Florida: Harcourt, Brace and Co., 1998.

Davidson, M. *Helen Keller's Teacher,* Scholastic Book Services, 1965.

_____. *Helen Keller,* Scholastic Book Services, 1969.

Glasby, P. "Recognizing What Is Human in the Practice of Education," *Journal for Waldorf/Steiner School Teachers* 3.2, 2001.

Goethe, J.W.v. "Der Versuch als Vermittler von Objekt und Subjekt" in *Naturwissenschaftliche Schriften. Goethe's Werke, Band 13,* Hamburg: Christian Wegner Verlag, 1960.

Guttenhöfer, P. "The Division of the Main Lesson and the Role of the Conclusion," *Journal for Waldorf/Steiner School Teachers* 6.1, 2004.

Maier, G. *Optik der Bilder,* Dürnau: Verlag der Kooperative, 1986.

Steiner, Rudolf. *Occult Science,* 1925, Chapter 2, "Sleep and Death," London: Steiner Press, 1969.

_____. *The Study of Man,* 1919 (Now translated as *Foundations of Human Experience*), Lectures 8 and 9; London: Rudolf Steiner Press, 1966.

_____. *Supplementary Course–the Upper School,* 1921 (also translated as *Education for Adolescents*), Lecture 3. Authorized translation by permission of the Rudolf Steiner Nachlassverwaltung, Switzerland. Translation issued by the College of Teachers, Michael Hall, Forest Row, Sussex, 1965.

_____. *The World of the Senses and the World of the Spirit,* six lectures given in Hanover, December 27, 1911–January 1, 1912, London: Rudolf Steiner Publishing Co., 1947.

Von Mackensen, M. *Klang, Helligkeit und Wärme,* Kassel, Germany: Bildungswerk, Brabanter Strasse 43, 3500, 1992.

Peter Glasby is a teacher of physical sciences at the Mt. Barker Waldorf School in Australia. He is highly regarded worldwide for his knowledge of phenomenology.

Thoughts on Information and Communication Technology

by Florian Oswald
Rudolf Steiner School, Bern, Switzerland
Translation by Diederic Ruarus and Martin Gastinger

As parents, teachers, and educators—in short, as contemporaries, we are confronted daily with the achievements of technology. We are challenged continuously to find a relationship with technology. In that arena information and communication technology holds a key position. I would like to try to bring some thoughts together with respect to this theme and develop some ideas about how to use this technology. Two pillars, which support a sensible usage, can be described thusly:

1. Technology can always be misused. We can try to understand the apparatus and thereby bring about consciousness of its functioning.

2. Through our own inner steadfastness we can evolve sensible use of technology: ethical, wise, safe.

To elaborate on what I mean in the first pillar, I would like to walk through the history and development of the different media. The first mass medium was the book. But only printing made it that. Prior to the printing press, books were the privilege of an elite group—there were so few and only a few individuals knew how to read. The importance of printing in transforming the book into the first mass media has often been underestimated.

A next milestone was the development of photography (1839). It was developed originally so that an image of a building could be preserved as a document for posterity. Photography can capture an object. A collection of accurate images can be made. With the medium of photography, one has the capacity to remove a particular detail from a field of vision and store it, first onto a (photographic) plate, later onto paper or another printable surface. A three-dimensional object can be recorded in two dimensions (on a plane). An anecdote from Picasso can clarify this. A lady was annoyed by the way Picasso painted people and asked him: "Can't you paint a person as he appears to us?" Picasso looked at her with astonishment and asked her what she meant by that. The lady produced a photograph of her husband and explained to Picasso that this was a proper image of him. After a short glance at the image, Picasso said: "I hadn't thought that your husband was so flat."

The telegraph opened the communications door a little further. In his book, *Amusing Ourselves to Death*, Neil Postman wrote:

> The strength of the telegraph was to transmit information, not in storing, explaining or analyzing it. In this sense the telegraph is the exact opposite of book printing. Books for example are excellent containers to store, for calm sifting and systematic analysis of information and ideas. Writing and reading a book take time.

With the telegraph, the transport of news was no longer dependent on the speed a human could travel, but on that of electricity. We should keep in mind, however, that the receiver does not get the original.

Printing, photography and telegraphy were all put to use by the newspaper. Increasingly, news items from all over the world were in circulation and photos as accompaniment proliferated. Over time the newspaper has changed from being a news medium to a medium of entertainment.

In 1876 the first telephone conversation took place—a conversation with a non-present person held in real time. We should keep in mind, however, that the voice of the non-present person is an imitation. Edison also thought about how to store the human voice, and in 1877 he unveiled the phonograph. Initially purely mechanical, the machine could play a piece of music in exactly the same rendering as the original and as often as we wanted to hear it. In the case of live music, however, the same piece sounds different every time because it is played slightly differently every time.

Again Edison led the way in the transition from photography to moving images. In his application for a patent (October 18, 1888), he wrote: "I am working on an instrument that does for the eye what the phonograph did for the ear, namely the recording and reproduction of moving objects, and that in a cheap, practical and simple manner." The medium of film allowed for bringing various elements together. The first film to show a coherent story was *The Life of an American Fireman*. In this documentary film, scenes of a burning house were interspersed with scenes from a theatre performance, both events having been recorded at different times and in different locations. The audience saw a coherent picture and could not discern that two completely distinct shootings had been combined.

A new era in information and communications was ushered in with the introduction of electronic media: radio, television and computer. Radio today does not play the same dominant role as it did when it was first introduced, when national radio was the news dispenser purely and simply. A colorful mix of information, radio dramas and music was the next development. Today most young people listen only to music and advertisements.

In its beginnings, television approached moving pictures and sound quite differently than did the cinema. Cinema meant films, and still does. Television began as an electronic newspaper. By watching TV, one could see and hear up-to-date national and international news. Today television offers such a flood of information, often hard to digest, and entertainment of questionable veracity and value. In particular we need to make a tremendous effort to establish a link between the images and reality.

According to Neil Postman, the problem with television is not that it brings us entertaining topics; the problem is that every topic is presented as entertainment. The use of television also ushered in major concerns, discussions, and studies in many areas of society about both the content and the time spent in front of the television, resulting in an extensive body of literature demanding a more conscious use of this medium.

The invention of both audio- and videotape have allowed for the listening and viewing habits of media to become more individualized and isolated. One can put his own program together according to personal wishes, needs, and schedule. And by using a video camera and video player, one can view his own pictures without even having to process film.

These media are all task-specific and specialized in their focus and function, i.e., there was not an instrument that could integrate or mix the technologies and provide all the information of seeing, hearing and speaking all in one. Until the computer. The computer is the first technology that attempts it all. Computer uses are not just determined by its construction but are ever-expanding through programming, software. Human ingenuity has turned the computer to a multitude of services: writing, typesetting, drawing, painting, photography, film, research, composing music, printing, copying, data sorting, creating databanks, storing, calculating, learning, programming, driving appliances, listening to radio, watching television and video, CDs, iPODS, games, communicating; image telephoning, e-mail, chat rooms; online services such as shopping, banking, trading, advertising, house-hunting, pornography, and so forth, and "virtual" everything and anything.

The list goes on and on. The broad range of its applications continues to grow and expand—ever-new opportunities targeting every age group. There is a special appeal to the virtual worlds, which can be created, explored, even believed. Everything seems possible. Here, the user is not in the traditional passive role as when using television, but one can contribute to, even "join in," the action on the screen.

This brief excursion into the world of media demonstrates that, over the course of its development, more and more senses have become engaged and involved. It is only a question of time before all our senses can be reached by cyberspace. With the introduction of electric and electronic media, the arts have increasingly been included. First music, then drama, and now almost every artistic aspect can be imitated or created by the computer. It is now possible to produce and reproduce art in many ways and with a fairly high degree of quality, raising the questions more and more frequently about what is art, what is aesthetics, and if what we can create on the computer is art.

How do we respond to the different technological achievements? What kind of interactions take place? For every technology a special consciousness, a certain maturity is required. The devices of media described above have changed our skills and skill levels and present challenges to us. In its time, book-printing brought about a far-reaching change: Handwriting was replaced by print, a process which today we take for granted! It is characteristic of such processes that we get used to them. But what are we giving up or, worse, losing? A talk on the phone and a face-to-face conversation are qualitatively different. What is appropriate to discuss in a phone call and what kind of conversations deserve or require personal connection? Often it is not easy to get away from the television; programming literally draws us in or entices us to surf the channels. Most music-listening today is through speakers that surround us or on devices that can reproduce only partials of the originals. The computer has found its place in daily chores and often determines the way they are carried out. A long list of observations from daily life could be included here.

We live in a high tech world. Individual differences, talents and strengths are disappearing through the application of technology. Strength of character is required. Are we still able to say no and decide freely which technology we apply, to which situation and at which time? Both school and home are faced with a challenge: to build and support the second pillar. Three areas of importance need to be considered and developed: the senses, the ability to discern, and inner strength.

All media address and engage the senses, but each in a one-sided, therefore incomplete, way. There's always something missing. In a phone call the holistic perception of the speaker is missing; he/she is reduced to voice only. Television appeals both to vision and to hearing; one might assume the experience to be more complete; but the screen is flat, three dimensions are reduced to two, and the lower senses are not touched. For the human being, to establish a real "connectedness" with the world, with matter, is a fundamental skill. This skill helps us to build a healthy relationship with our physical body. An intensive training of the senses in the first and second years of life can provide security.

With forming a judgment, it is the same as with many things that require a certain maturity. If influences or demands come too early, we will not have the opportunity to find out how a higher level of maturity would have managed or what could have been accomplished. We can learn to discern things and ask ourselves whether it makes a difference to listen to a live concert or to music from a loudspeaker. Does it matter whether I type a letter on the computer or write it by hand? What kind of relationship do I have to the images and sounds transmitted by television, to the feelings these images evoke, to the knowledge or disinformation they bring?

A huge task lies ahead if we are not to be taken over by technology. A gesture of distancing oneself is always necessary and subsequently a finding of what is appropriate in each situation. As we cannot know everything, we rely on authorities. The skill we need to acquire then is the ability to discern truth. We can learn to discriminate who is a good judge and who is not. The central issue is how to nurture the faculty of judgment. Inner steadfastness is another key issue. I observe in myself and others how these media devices have a tendency to possess us, to suck us in. I lose track of time in front of the computer, lose the will to move away from the television because I want to see what's next. I feel the security of being always available by cell phone.

The use of media technology requires inner discipline from me, a self-affirmation in the truest sense of the word. In short, I must school myself to be able to say no. I am called to choose, to take responsibility for my actions. This requires strengthening and training of the will. Here the toddlers can teach us a lot. Toddlers are persistent and ready to master all kind of obstacles. In contrast, the world of adults and adolescents is suffering from something that these younger human beings do not know at all: laziness. Our culture tends to remove all obstacles for our young people! Experience shows however that the stones we remove from their path they tend to throw at us later. Our task is not to remove the stones but to help the young people to find ways to deal with them.

In the presence of modern technology there lies a wakeup call of a special kind. Pure thinking is not enough. There must be will activity. For example, I can learn at the desk how a car functions and all its parts, but I still cannot drive it until I do it. Thought-filled *intellectual understanding* needs to be complemented by thought-filled *application*, because without application the device remains only an idea. Today the opposite is the case. We apply technology, which we do not always understand, totally caught up in the action. Neither experience alone nor idea alone will lead us to a responsible way of dealing with technology. The healthy insight comes from the meeting of those two poles. In this sense, modern technology is a challenge that we have to tackle consciously. We are risking a lot, but by taking hold of ourselves, we stand to gain a lot.

Florian Oswald teaches mathematics and physics, among other subjects. He is a member of the Hague Circle which organizes the International Conference which takes place in Dornach, Switzerland, every four years. He has made warm connections with the anthroposophical work in the southern hemisphere over some years.